Fundamentals of Smart Materials

Print ISBN: 978-1-78262-645-9
EPUB ISBN: 978-1-78801-946-0

A catalogue record for this book is available from the British Library

The Royal Society of Chemistry is a charity, registered in England and Wales, Number 207890, and a company incorporated in England by Royal Charter (Registered No. RC000524), registered office: Burlington House, Piccadilly, London W1J 0BA, UK, Telephone: +44 (0) 20 7437 8656.

Visit our website at www.rsc.org/books

Printed in the United Kingdom by CPI Group (UK) Ltd, Croydon, CR0 4YY, UK

Fundamentals of Smart Materials

Edited by

Mohsen Shahinpoor

University of Maine, USA
Email: shah@maine.edu

ROYAL SOCIETY
OF **CHEMISTRY**

Preface

There have been tremendous efforts put into developing smart materials as stimuli-responsive multifunctional materials with actuation, energy-harvesting, and sensing capabilities, as well as companion thermal, electromagnetic, and chemical functions. There are only a few textbooks available, which only cover a shortlist of stimuli-responsive smart materials. These existing textbooks have been partially reviewed in various chapters of this book in the context of materials science, chemistry, medicine, physics, or engineering.

Smart materials appear to be playing a fundamental role in the advancement of engineering, science, technology, and medicine. This textbook intended for seniors and graduate students briefly reviews the fundamentals of more than twenty-five families of smart materials, defined as materials with some actuation, energy harvesting, and sensing capabilities among other chemical, fluid, thermal or electromagnetic properties. Today, many researchers around the world are actively involved in searching for new smart materials.

In what follows, the scene will be set for showing the importance and significance of multi-functional stimuli-responsive smart materials, particularly with some actuation, energy harvesting, and sensing characteristics and some additional companion properties such as sensing thermal, fluid, chemical, physical and solid mechanical interactions and be capable of producing such fields in a reverse manner.

Senior undergraduate and graduate students, majoring in engineering, materials chemistry, biomedical engineering, medicine and materials science, as well as chemistry and physics, tested the content of the present book, which tries to cover as many as possible smart materials families. This education is reflected in the book content, which covers a broad range of topics, from piezoelectrics to self-healing. Being

Fundamentals of Smart Materials
Edited by Mohsen Shahinpoor
© The Royal Society of Chemistry 2020
Published by the Royal Society of Chemistry, www.rsc.org

stimuli responsive, smart materials typically require an input, as a trigger to execute the instructions and perform the desired functions.

In essence, the main objective of this book is to provide comprehensive and concise reviews of over twenty-five families of stimuli-responsive smart materials. In particular, piezoelectric materials, piezoresistive materials, electrostrictive materials, fibrous polyacrylonitrile (PAN) materials as linear fibrous artificial muscles, giant magnetostrictive materials (GMS), giant magnetoresistive materials (GMRs), magnetic gels (MGs), electrorheological fluids (ERFs), magnetorheological fluids (MRFs), dielectric elastomers (DEs), shape–memory alloys (SMAs), magnetic shape–memory alloys (MSMs), shape–memory polymers (SMPs), smart materials for controlled drug release, mechanochromic and metamaterials, ionic polymer–metal nano composites (IPMCs), smart ionic liquids (ILs), conductive polymers (CPs), liquid crystals (LCs) and liquid crystal elastomer (LCEs), chemo-mechanical polymers and smart nanogels for biomedical applications, smart self-healing materials and smart Janus particles are reviewed.

Earlier versions of this book were assigned as textbooks for the editor's Smart Materials graduate course and went through many iterations. The editor is very grateful to all of the University of Maine students who used the content of this book and provided useful feedback and insight.

The chapters are arranged such that there is an introduction to the subject followed by a description of the properties of the materials and their applications with example problems followed by brief modeling of the constitutive equations, a brief discussion on fabrication and manufacturing, followed by conclusions and references. There are 24 chapters, all of include exercises and homework problems. An important educational, **Appendix A**, is a listing of the short synopses of books on smart materials published by the Royal Society of Chemistry. This list of materials in **Appendix A** is numerically greater than the 25 families of the smart materials covered in this textbook.

In the chapters that follow, the importance and significance of multi-functional smart materials, particularly with some actuation, energy harvesting, and sensing characteristics, as well as some additional companion properties, such as sensing, chemical, thermal, fluid and solid mechanical interactions and being capable of producing such fields in a reverse manner, will be shown.

Mohsen Shahinpoor
University of Maine

Acknowledgements

The editor would like to thank a number of individuals who greatly helped the creation of this textbook on the fundamentals of smart materials. Sincere thanks and appreciation are extended to Professor Hans-Jörg Schneider, FR Organische Chemie, Universität des Saarlandes, Germany, for all his mentorship and guidance throughout the preparation of this textbook. Thanks are also extended to the Royal Society of Commissioning Editors Dr Leanne Marle, MRSC, and Dr Robin Driscoll for all their efforts in promoting, expanding and reviewing the Smart Materials Series, as well as the current textbook on the *Fundamentals of Smart Materials*. Thanks are also extended to the Royal Society of Chemistry for editorial support from Connor Sheppard for great help towards reviewing and revising the manuscripts of various chapters. Last but not least, our thanks are extended to the Royal Society of Chemistry for the publication of the Smart Materials Series, as well as this textbook *Fundamentals of Smart Materials*.

Mohsen Shahinpoor

Fundamentals of Smart Materials
Edited by Mohsen Shahinpoor
© The Royal Society of Chemistry 2020
Published by the Royal Society of Chemistry, www.rsc.org

Contents

Fundamentals of Smart Materials
Edited by Mohsen Shahinpoor
© The Royal Society of Chemistry 2020
Published by the Royal Society of Chemistry, www.rsc.org

4 Review of Electrostrictive Materials   36
Mohsen Shahinpoor

5 Review of the use of Fibrous Contractile Ionic Polyacrylonitrile (PAN) in Smart Materials and Artificial Muscles   46
Mohsen Shahinpoor

6   Review of Magnetostrictive (MSMs) and Giant Magnetostrictive Materials (GMSs)   64

Mohsen Shahinpoor

7   Review of Giant Magnetoresistive (GMR) Materials   73

Mohsen Shahinpoor

8   Review of Magnetic Gels as Smart Materials   84

Mohsen Shahinpoor

9   Review of Electrorheological Fluids (ERFs) as Smart Material   98

Mohsen Shahinpoor

10   Review of Magnetorheological Fluids as Smart Materials   107

Norman M. Wereley and Young Choi

11 Review of Dielectric Elastomers (DEs) as Smart Materials   118

Mohsen Shahinpoor

12 Review of Shape Memory Alloys (SMAs) as Smart Materials   136

Mohsen Shahinpoor

13 Review of Magnetic Shape Memory Smart Materials   151

Mohsen Shahinpoor

14 Shape Memory Polymers (SMPs) as Smart Materials   160

Mohsen Shahinpoor

15 Review of Smart Materials for Controlled Drug Release 170

Carmen Alvarez-Lorenzo and Angel Concheiro

16 Review of Smart Mechanochromic and Metamaterials 193

Mohsen Shahinpoor

Subject Index   322

1 General Introduction to Smart Materials

Mohsen Shahinpoor

Mechanical Engineering Dept., University of Maine, USA
Email: shah@maine.edu

1.1 Introduction

Before elaborating on the nature of this textbook, let us redefine smart materials as stimuli-responsive multifunctional materials with actuation, energy-harvesting, and sensing capabilities, along with companion thermal, electromagnetic, chemical, and physical functions. As mentioned in the preface, there are only a few smart materials textbooks available, which only cover a shortlist of stimuli-responsive smart materials. The current textbook reviews some 25 families of smart materials.

In this introductory chapter, the spirit and structure of this book are presented. In general, some fundamental aspects of various smart materials are described, and the stage is set for the coverage of the current family of smart materials as special stimuli-responsive smart materials capable of a variety of actual functions needed for a large family of engineering, scientific, industrial and medical applications. These functions include actuation, energy harvesting, and sensing, plus some other complimentary physical or chemical properties changed *via* external or internal stimuli such as electric or magnetic fields, fluid-thermal fields, strain and stress fields, plus others such as the ionic field within the materials.

Fundamentals of Smart Materials
Edited by Mohsen Shahinpoor
© The Royal Society of Chemistry 2020
Published by the Royal Society of Chemistry, www.rsc.org

The chapters are presented such that initially an Introduction is given followed by applications with example problems, followed by brief Modeling on Constitutive Equations, a brief discussion on Fabrication and Manufacturing followed by Conclusions and References where 'properties' is used in the broad sense of the word. 'Stimuli-responsive' or 'Multi-functional' could cover just about any material, but here these terms are meant as special functions, such as actuation, energy harvesting, and sensing, among others. The chapters are also equipped with exercises and solutions, as well as homework problems. The book also has a solutions manual for the homework problems, available as ESI.[†] This book will have twenty four chapters, including this chapter, which is an general introductory chapter on the fundamentals of smart materials. Each chapter will describe the characteristics of a particular material and system that is currently available and can be manufactured or fabricated to act as stimuli-responsive multi-functional smart actuators, sensors, and energy harvesters, among other functions and properties.

Let us briefly present a summary of the book to prepare the students for more comprehensive coverage of the materials and topics. Because of their long history, a review on piezoelectric materials, such as piezoceramics like PZT and piezo polymers like PVDF is presented. The piezoelectric effect describes a reversible electrodynamic relationship that exists in some solid crystalline structures with embedded dipoles. These crystalline solids possess microscopic regions containing dipole charges such that when placed under an applied mechanical stress, they change their internal arrangements of embedded electrical dipoles, generating a voltage across the material boundaries. Conversely, an applied voltage to the solid crystalline changes the orientation of the embedded internal dipole charges and generates deformation or strain in the solid. The description of piezoelectric materials is then followed by a review on piezoresistive materials as smart sensors. Piezoresistivity is a property of certain materials such as semiconductors for which the materials electrical resistance changes purely due to mechanical pressure, force, acceleration, strain, and stress. It is the physical property of certain materials that have been widely used to convert a mechanical signal into an electrical signal, in smart sensors, accelerometers, tactile sensors, strain gauges, flow meters, and similar devices and microdevices.

[†]Electronic supplementary information (ESI) available. See https://pubs.rsc.org/en/content/ebook/978-1-78262-645-9.

The piezoresistive effect is present in semiconductors such as germanium, amorphous silicon, polycrystalline silicon, silicon carbide, among other materials.

A concise review on the response of electrostrictive materials is then given. Electrostriction is the nonlinear electromechanical coupling in all electrical-nonconductors (dielectric materials). Under the application of an electric field, these materials show deformation, strain, and stress. Generally speaking, all electrostrictive materials exhibit second-order nonlinear coupling between the elastic strains or stresses and dielectric terms, such as the strain tensor. For a single uniaxial strain (deformation), the induced strain (deformation) is directly proportional to the square of the applied electric field (voltage).

Contractile ionic polyacrylonitrile (PAN) fibers are then introduced to mimic mammalian muscles. Polyacrylonitrile (PAN) fibers in an active form (PAN or PAN gel modified by annealing/cross-linking and partial hydrolysis) elongate and contract when immersed in pH solutions (caustic and acidic solutions, respectively). Activated polyacrylonitrile (PAN) fibers can also contract and expand in polyelectrolyte when electrically and ionically activated with cations and anions, respectively. The change in length for these pH-activated fibers is typically greater than 100%. However, more than 900% contraction/expansion of PAN nanofibers (less than 1 micron in diameter) have been observed in our laboratories. PAN muscles present great potential as artificial muscles for linear actuation, and PAN fibers can convert chemical energy directly into mechanical motion.

Magnetostrictive materials are then introduced, in which deformation is observed in ferromagnetic materials when they are subjected to a magnetic field. This effect was first identified in 1842 by James Joule when observing a sample of nickel (the Joule Effect). At the fundamental level, the change in dimensions results from the interactive coupling between an applied magnetic field and the magnetization and magnetic moments of the material's magnetic dipoles, for a material initially under some stress. A review of giant magnetoresistive (GMR) materials is then presented. Magnetoresistance is defined as the property of a material whereby it can change its electrical conductivity or inverse electrical resistance when an external magnetic field is applied. In 1851, William Thomson (Lord Kelvin) discovered that when pieces of iron or nickel are placed within an external magnetic field that the electrical resistance increases when the current is in the same direction as the magnetic force which is aligned with the magnetic N–S vector and decreases when the current is perpendicular to the direction of the magnetic force. Lord Kelvin was unable to reduce the electrical resistance of any metal by more than about 5%. This effect is commonly called the ordinary magnetoresistance (OMR) effect to differentiate it from the more recent discovery of GMR. GMR materials generally possess alternating layers of ferromagnetic and non-magnetic but conductive layers made up of iron–chromium and cobalt–copper. Following the information on GMR, a brief review of magnetic gels (ferrogels) is presented by Zrinyi and co-workers. A prelude to the development of ferrogels was a classic paper by Rosenzweig in 1985 on ferrohydrodynamics. A colloidal ferrofluid, or a magnetic fluid, is a colloidal dispersion of monodomain magnetic particles. Typically, monodomain magnetic particles have typical sizes of around 10–15 nm, and they are superparamagnetic, in which magnetization can randomly flip direction under the influence of temperature.

A review of electrorheological fluids (ERFs) is then presented. ERFs belong to a class of smart materials capable of changing from a liquid phase to a much more viscous liquid and then to an almost solid phase in the presence of a dynamic electric field. They are essentially colloidal suspensions of highly polarizable particles in a nonpolarizable solvent. The solid phase of an ERF typically has mechanical properties similar to a solid like a gel and can undergo a phase change from liquid to a thick liquid like honey and then solid or in reverse from a solid transform to a thick liquid and then a thin liquid in a matter of a few milliseconds. This effect is called the "Winslow effect" after its discoverer Willis M. Winslow, who obtained a US patent on the effect in 1947 and published an article on it in 1949. The effect is better described as electric field dependent shear yield stress. Magnetorheological fluids (MRFs) are then introduced, which are suspensions of micron-sized magnetic particles such as iron carbonyl powder in a host liquid, usually a type of oil with some additives, to minimize particle sedimentation and particle wear and tear. When the MRF suspension is placed in a magnetic field, the suspended colloidal particles reconfigure to form chains in the direction of the magnetic flux and make the solution more solid-like than liquid.

A review of dielectric elastomers (DEs) is then presented. If rubbery elastomers like a silicone rubber sheet are sandwiched between two compliant electrodes, then any imposed electric field induces electrostatic forces (attraction) between the electrodes. Thus, the rubber sheet in between them can be compressed by the electrostatic forces, which then causes the rubbery sheet to expand sideways due to the Poisson's ratio effect and actuation results. In 1880, Röntgen demonstrated this actuation using two glasses as dielectrics, and once the opposing surfaces of these glasses were charged, small thickness changes were observed. Later, electrostatically-induced pressures acting to compress dielectrics became known as the "Maxwell stress." It was, however, Pelrine, Kornbluh, and Joseph in 1998 who introduced dielectric elastomer technology with compliant electrodes. They concluded that by deliberately choosing polymers with relatively low moduli of elasticity, the field-induced strain response due to Maxwell stress could be large.

Shape-memory alloys (SMAs) are then reviewed. The shape-memory effect (SME) is a property of materials that are capable of solid-phase transformation from a body-centered tetragonal form called thermoelastic martensite to a face-centered cubic superelastic called austenite. These materials are named shape-memory materials (SMMs) and the thermal versions are called SMAs. These martensitic crystalline structures are capable of returning to their original shape in the austenite phase, after a large plastic deformation in the martensitic phase and return to their original shape when heated towards austenitic transformation. These novel effects are called thermal shape-memory and superelasticity (elastic shape-memory), respectively. Magnetic shape-memory (MSM) alloys (materials) are then described. MSMs, often also referred to as ferromagnetic shape-memory alloys (FSMAs), have emerged as an interesting extension of the class of SMMs. FSMAs combine the attributes and properties of ferromagnetism with a reversible martensitic crystalline solid phase transformation. MSM phenomena were originally suggested by Ullakko, O'Handley, and Kantner and were demonstrated for a Ni–Mn–Ga alloy in as early as 1996. Naturally, the SME is now extended to polymers as shape-memory polymers (SMPs). SMPs belong to the family of SMMs, and are stimuli-sensitive polymers that can be deformed into a predetermined shape under some specific applied fields or parameters such as temperature, electric or magnetic field, as well as strain and stress. These shapes can be relaxed back to their original field-free shapes under thermal, electrical, magnetic, strain, stress, temperature, laser, or environmental stimuli. These transformations are essentially as a result of the elastic energy stored in SMMs during the initial deformation.

Smart materials for controlled drug release are then described. Systemically-administered controlled release systems allow fine-tuning of drug bioavailability, *via* regulation of the amount and rate at which the drug reaches the bloodstream, which is critical for the success of the therapy. Some drugs pose important efficacy and safety problems (*e.g.*, antitumor drugs, antimicrobials) and suffer from in-stability problems in the biological environment (*e.g.*, gene materials), and thus the therapeutic performance of these drugs is improved when they are selectively directed (targeted) from the bloodstream to the site of action (tissues, cells or cellular structures). Both macro-dosage forms and nano-delivery systems may notably benefit from stimuli-responsive materials. Differently, to pre-programmed drug release systems, formulations that provide discontinuous release as a function of specific signals (stimuli) are advantageous in many situations. The when, where and how drug release triggering occur require detailed knowledge of the changes that the illness causes, in terms of physiological parameters. These changes can be character-ized in terms of biomarkers (*e.g.*, glucose, specific enzymes, or *quorum sensing* signals in the case of infection) and physicochemical par-ameters (pH, ions, temperature, glutathione) that may be exploited as internal stimuli. When the physio-pathological changes are too weak or non-specific, the application of external stimuli may be an alternative. External sources of temperature, ultrasound, light, and magnetic or electric fields may allow for the focal switch on/off of drug release.

Mechanochromic smart materials, as well as mechanical metama-terials, are then reviewed. In particular, two recent families of smart materials, namely mechanochromic materials and mechanical me-tamaterials, respectively, are described. Mechanochromic materials change their optical properties, and in particular, photoluminescence characteristics, if subjected to mechanical loading. Metamaterials are materials that are not ordinarily produced in nature. Smaller units rather than the properties of the host material play a fundamental role in materials behavior. Metamaterials are nanocomposite materials made up of periodically repeated micro or nano units of metals, alloys, and plastics that exhibit properties different from the natural properties of the participating materials.

Ionic polymer–metal composites (IPMCs) are then reviewed. Ionic polymeric networks contain conjugated ions that can be redistributed by an imposed electric field and consequently act as distributed nano actuators, nanosensors, and energy harvesters. This chapter briefly presents the manufacturing methodologies and fundamental properties and characteristics of ionic polymers such as IPMCs. Gel-based and chitosan-based conductor composites have also been considered as electrically active composite smart materials similar to IPMCs. Following the description of IPMCs, a review of smart ionic liquids is then presented. Ionic solids such as sodium chloride (table salt) have been known for centuries. The very first examples were the groundbreaking endeavors of Sir Humphry Davy in the synthesis of alkali metals by electrolysis. However, this process needs to be carried out at a high temperature as the ionic bonds are strong. Electrolysis of sodium chloride should therefore be conducted at a temperature higher than 801 °C. Since high temperatures are not technologically favorable, the melting points of such ionic solids are reduced by weakening the ionic bonds in eutectic mixtures. One of the very first examples of this was the pioneering work of Charles Martin Hall in the synthesis of aluminum, which is still the dominant approach for the exploitation of metallic aluminum. The high melting point of these ionic liquids is due to the close arrangement of highly charged ions within the lattice.

Conductive polymers are then reviewed. There are currently a fairly large number of conducting polymers or synthetic conductors that are being used industrially or medically. Some of the basic conducting polymers are polypyrrole, polyaniline, polythiophene, polyphenylvinilene, and polyacetylene, which can be manufactured *via* chemical or electrochemical oxidation and reduction (REDOX) procedures. Conductive polymers with the ability to conduct electrical charges in addition to being flexible, optically active and not difficult to synthesize present a tremendous opportunity for the industrial and medical applications of conductive polymers. Pioneering work on conductive polymers reported the observation that the conductivity of polyacetylene increases by millions of times when it is oxidized by "doping" with iodine vapor. Conductive polymers can conduct electrical charge because within their molecular network, charges can jump between the molecular chains of the polymer. Conductive polymer molecular structures possess both single and double chemical bonds, which enhance charge transfer.

A description of liquid crystal elastomers, as smart materials, is then presented, a topic first discussed by Finkelmann. These materials can be used as robotic actuators through the induction of a nematic–isotropic phase transition within them upon an increase in temperature, which causes them to shrink, as described in a review on these intelligent multi-functional materials by Brand and Finkelmann. LCEs have been made electroactive by creating a composite material that consists of nematic LCEs and a conductive phase such as graphite or conducting polymers that are distributed within their network structure. The actuation mechanism of these materials involves a phase transition between a nematic (cholesteric, smectic) and isotropic phases over less than a second. The reverse process is slower, taking about 10 seconds, and requires cooling of the LCE back to its initial temperature as the LCE expands back to its original size.

Chemoresponsive gels as smart materials are then discussed. Hydrogels can be used for many applications, and are being increasingly developed given their possible biocompatibility. Such smart materials can, depending on suitable chemical components, bind or release, for example, drugs, pollutants, catalysts, *etc.*, upon interaction with external effectors, and swell or shrink under the influence of different pH, various chemical compounds, temperature, or light. Most hydrogels are amorphous, some are semicrystalline mixtures of amorphous and crystalline phases, or are crystalline. Hydrogels have a water content typically between 80–99%, which can be changed by external stimuli; this is the basis of many applications. Natural sources of hydrogels are, for example, agarose, chitosan, methylcellulose or hyaluronic acid, but most smart hydrogels are based on synthetic polymers or rely on chemical modification of natural systems. Synthetic polymers for gels are usually obtained *via* copolymerization or cross-linking free-radical polymerization, where hydrophilic monomers are reacted with multifunctional cross-linkers. One can produce polymer chains by chemical reaction, *via* photochemical processes, or using radiation for the generation of free radicals. Alternatively, one can modify existing polymers through chemical reactions.

Smart nanogels for biomedical applications are then described. Smart nanogels are one of the most important innovations that have emerged in the field of nanomedicine and biomedical applications. Through recent advances in the applications of biomaterials, nanogels have emerged as novel candidates for drug delivery, biosensing, imaging, tissue engineering, and the targeted delivery of bioactive compounds. The present chapter gives a basic understanding of hydrogels and introduces the nanoparticle form of hydrogels known as "nanogels." Nanogels have synergistic properties of the interpenetrating networks as well as nanoscale properties, such as small size and high surface-to-volume ratio. These hybrid materials show high drug loading, are capable of crossing strong barriers, and are also highly biocompatible. In brief, this chapter describes the basic synthetic methodology and characterization techniques of nanogels. It also discusses the natural and synthetic polymers deployed for the synthesis of nanogels.

Self-healing materials are then introduced and discussed. The self-healing characteristics of these materials and in particular biomaterials and the concepts of the self-healing processes in nature and biology are already well known by scientific communities. One can start by describing their impact and occurrence in nature, in plants, in animals and human beings. These understandings of self-healing processes in biology and nature are particularly more advanced in terms of dermatology and skin repair by scar tissues. The advantage of self-healing materials is that they can treat material degradation by initiating a repair mechanism that responds to the incurred damage or degradation.

Finally, a review of Janus particles as smart materials is presented. In ancient Roman times, *Janus* was the god who had two faces (beginnings and endings). In modern science, we have adopted the term to describe particles with two distinct and usually contrasting sides. These particles have the resemblance of the Taijitu symbol in ancient Asian philosophy, where Yin and Yang (dark and bright) were used to describing seemingly opposite forces. It is believed that these two basic elements give rise to complicated change and transition in the whole world. In the same sense, Janus particles are defined by their duality, which can take on a variety of forms and create a wide range of new materials with the simple Janus motif. The possibilities for properties that can be assigned to each half of the Janus particles are vast (for example, hydrophobicity and charge), and are limited only by the fabrication capabilities of their creators.

Homework Problems

For each of the following materials, provide a simple description of their actuation and sensing characteristics. Are they just actuators, sensors or both?

Homework Problem 1.1

Piezoelectric materials

Homework Problem 1.2

Piezoresistive materials

Homework Problem 1.3

Electrostrictive materials

Homework Problem 1.4

Fibrous polyacrylonitrile (PAN) artificial muscles

Homework Problem 1.5

Giant magnetostrictive materials

Homework Problem 1.6

Giant magnetoresistive materials (GMRs)

Homework Problem 1.7

Magnetic gels

Homework Problem 1.8

Electrorheological fluids

Homework Problem 1.9

Magnetorheological fluids

Homework Problem 1.10

Dielectric elastomers

Homework Problem 1.11

Shape-memory alloys

Homework Problem 1.12

Magnetic shape-memory alloys

Homework Problem 1.13

Shape-memory polymers

Homework Problem 1.14

Mechanochromic and metamaterials

Homework Problem 1.15

Ionic polymer–metal nano composites

Homework Problem 1.16

Smart ionic liquids

Homework Problem 1.17

Conductive polymers

Homework Problem 1.18

Liquid crystals and liquid crystal elastomers

Homework Problem 1.19

Chemomechanical polymers

Homework Problem 1.20

Nanogels

Homework Problem 1.21

Self-healing materials

Homework Problem 1.22

Janus particles as smart materials

Abbreviations and Acronyms

PZT	Lead zirconate titanate
PVDF	Polyvinylidene difluoride
PMN	Lead magnesium niobate
MSMs	Magnetostrictive materials
GMRs	Giant magnetoresistive materials
ERFs	Electrorheological fluids
MRFs	Magnetorheological fluids
EAPs	Electroactive polymers

DEs	Dielectric elastomers
IPMCs	Ionic polymer–metal composites
IPMNCs	Ionic polymer–metal nanocomposites
CPs	Conductive and conjugated polymers
MHAMs	Metal hydride artificial muscles
PAN	Polyacrylonitrile
LCEs	Liquid crystal elastomers
SMAs	Shape-memory alloys
Nitinol	Nickel–titanium–NOL
MSM	Magnetic shape-memory
SMPs	Shape-memory polymers
SHMs	Self-healing materials
REDOX	Oxidation–reduction procedure

2 Review of Piezoelectric Materials

Mohsen Shahinpoor

Mechanical Engineering Dept., University of Maine, USA
Email: shah@maine.edu

2.1 Introduction

2.1.1 The Piezoelectric Effect

The piezoelectric effect describes the reversible electrodynamic relationship that exists in some solid structures with a non-centrosymmetric crystalline structure containing embedded dipoles. These crystalline solids possess microscopic regions containing dipole charges such that under applied mechanical stress the internal arrangements and orientations of the embedded electrical dipoles change, subsequently generating a voltage across the material boundaries. Conversely, a voltage applied to the solid crystalline changes the orientation of the embedded internal dipole charges and generates a deformation or strain in the solid. Figure 2.1 below displays the piezoelectric effect on a solid cube in (a) neutral, (b) expanded and (c) contracted configurations in the presence of imposed electric or deformation fields. Note that the polarization direction of poling of the internal dipoles is shown as the P arrow.

Brothers Jacques and Pierre Curie discovered the piezoelectric effect in 1880 while working on quartz.[1-3] The word "piezo" originates

Fundamentals of Smart Materials
Edited by Mohsen Shahinpoor
© The Royal Society of Chemistry 2020
Published by the Royal Society of Chemistry, www.rsc.org

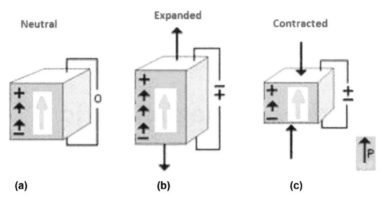

Figure 2.1   Deformation of a cubic piezoelectric solid (poled, P) under no pressure or stress (neutral, a), under an electric or deformation field in the polarization direction (expanded, b) and under an electric or deformation field opposite to the polarization direction (contracted, c).

from the Greek word "piezein", meaning press or pressure. The Curie brothers observed that imposing pressure or a deformation field on several crystals such as quartz (SiO_2) and tourmaline quartz generate electrical charges. They called this phenomenon the "piezoelectric effect." They further discovered that this effect is reversible and noted that imposed electrical fields can generate deformations such as expansion and contraction in piezoelectric materials. This effect is called the "inverse piezoelectric effect".[4] In 1880, Rochelle salts and tourmaline were the best known piezoelectric materials at that time. The discovery of piezoelectric materials by the Curie brothers led to extensive research and materials development, leading to the present highly effective piezoelectric ceramic materials that have adjustable parameters.

2.1.2   Microscopic Piezoelectricity

As described before, the microscopic origin of the piezoelectric effect is the displacement and redistribution of ionic charges within a solid crystalline structure.[5–7] Piezoelectric materials produce an electric field when they are subjected to a mechanical strain field, and if an electric field is applied to piezoelectric materials, a strain field results. Thus, piezoelectric behavior can be realized in two distinct ways. The first is called the piezoelectric effect, when a piezoelectric material becomes electrically charged under strain or stress and can output a voltage. This property can be used in energy harvesting and sensing, which measure strain, strain rate, forces, pressure, and vibration. The second property of piezoelectric materials is that they exhibit the converse piezoelectric effect in the sense that the piezoelectric material becomes strained and stressed when placed in an electric field.

2.1.3 Piezoelectric Ceramics

Piezoelectric ceramics have found many commercial applications as electronic transducers. Piezoelectric ceramics were developed from high-quality barium–titanate ceramics used in capacitors. These ceramics are mostly in the form of oxide materials based on lead oxide, zirconate oxide, and titanate oxide. Piezoelectric ceramics include several ceramics such as lead zirconate titanate (PZT), lead metanio-bate (LMN), lead titanate (LT), and lead magnesium niobate (PMN), where PZTs have been the most extensively produced and employed materials.

The crystalline structure of a PZT is derived from the mineral perovskite ($CaTiO_3$). This structure is formed above the Curie temperature from regular oxygen octahedra with titanate and zirconium ions at the center. No polarization can be measured above the Curie temperature because the directions of the single dipoles are distributed such that the crystal structures of materials such as PZT become centrosymmetric and hence do not have any dipoles. An applied electric field is capable of aligning the embedded dipoles in the crystalline piezoelectric solid along the direction of the field. Upon thermal quenching and after the electric field has been removed, the polarization alignment is established.

2.1.4 Piezo Polymers

There are also polymers that exhibiting piezoelectricity. For example, polyvinylidene fluoride (PVDF) is a well-known polymer that exhibiting the piezoelectric effect. Other well-known piezo-polymers are P(VDFTrFe) PVDF or polyvinylidene fluoride trifluoroethylene and P(VDF TFe) and polyvinylidene fluoride tetrafluoroethylene and their copolymers.[8] Other piezo polymers are polyamides and polyurea, as well as liquid crystal elastomers and polymers. The origin of the piezoelectric effect in polymer chains (*i.e.,* for PVDF) is related to the positive charges on the H ions and the negative charges on the F ions. In the beta phase, these ions are arranged in such a way to generate dipoles along the length of the chains.

2.1.5   Polarization of Piezoelectric Materials and the Role of the Curie Temperature, T_c

Since the Curie brothers' initial findings, many other piezoelectric crystals have been discovered, researched, and extensively used. These include the ceramic materials discussed in the previous section 2.1.3, as well as polymeric materials such as PVDF, and piezo-bio polymeric materials such as hair, bones, crab and lobster shells, skin, wool, and animal horns.[4] Note that the piezoelectric effect occurs only in non-electrically conductive materials and can be induced in crystalline materials and ceramics such as alloys of lead, zirconium, and titanium (PZT), lithium and niobium ($LiNbO_3$) or lithium niobate. These solids are first heated up to their Curie temperature, T_c, in a very hot oil bath, at which the dipoles become thermally mobile and orientable in the direction of the imposed high electric field.

The Curie temperature, as defined for ferromagnetic materials, is also used in piezoelectric materials to describe the temperature above which the material loses its piezoelectric characteristics due to the spontaneous polarization of dipoles. Note, that in lead zirconate titanate (PZT) ceramics, the material crystalline structure is tetragonal below T_c and exhibits a displaced central cation leading to a net dipole moment in the material. Above T_c, an electric field of sufficient strength is applied in the desired direction, realigning the dipoles along the "poling" axis. After poling, the material is rapidly quenched for the dipoles to retain their directional preference in the electric field and thus, the material becomes a solid piezoelectric material (Figure 2.2). The imposed electric field causes a reorientation of the spontaneous polarization of dipoles. Simultaneously, if some

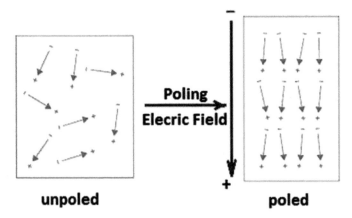

Figure 2.2   Poling and reorientation of internal dipoles by an applied electric field.

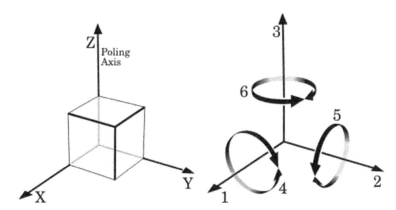

Figure 2.3 Right-handed coordinate system used in the piezoelectric effect, showing three axes of motion and three axes of rotation.

domains have a favorable orientation in line with the applied electric field these grow and expand, while domains with an unfavorable orientation with respect to the electric field shrink. Thus, the domain walls move within the crystalline lattice.

The piezoelectric axis and its relation to the poling axis are of special consideration. According to the IEEE Standard on Piezo-electricity,[9] the poling axis is denoted as the "3" axis or the "z" axis in a right-handed rectangular Cartesian coordinate system (x, y, z), as shown in Figure 2.3.

Before elaborating on the engineering models and systems for piezoelectricity, it should be emphasized that the piezoelectric effect is used in a large number of applications and products. They even extend to a refreshable Braille alphabet reader/writer for the blind. Piezo actuator and sensing technologies are also used extensively in the automotive technology for a large number of applications, such as sensing tire pressure to piezo-controlled fuel injection in combustion engines to speed up the transition times and help improve the quality of exhaust gas.

2.2   Piezoelectric Ceramic Actuators, Energy-harvesters, and Sensors

Piezoelectric ceramics have been extensively used as actuators, energy harvesters, and sensors in intelligent materials systems. Common servo-valve/hydraulic actuators suffer from various limitations, including multiple energy conversions, a large number of machine parts, vulnerability of the hydraulic pipe network, and frequency bandwidth limitations, according to Giurgiutiu, Qi *et al.* and Rafique and Bonello.[10–12] Note that piezoelectric ceramic actuators overcome these limitations by integrating with the host structure as embedded or bonded actuators. They further direct electrical energy to create high-frequency linear motions. When a voltage is applied to piezoelectric actuators, they undergo small displacements with a high force capability. Their key features include that they are compact and lightweight, and have a fast response, large force, displacements proportional to the applied voltage and broad operating temperature range. The most common piezoelectric ceramic actuators include ceramic patches, stacks, and piezoelectric fiber actuators.

2.3 Constitutive Modeling of Piezoelectric Materials

Piezoelectric ceramics can be modeled either numerically or analytically using basic constitutive equations. Non-linear modeling of piezoelectric ceramics has been discussed by Yousefi-Koma and Vukovich[13] and Chan and Hagood.[14] Tzou *et al.*[15] presented the mathematical modeling of a non-linear laminated anisotropic piezoelectric structure. Tzou *et al.*[15] further proposed a generic theory and derived its non-linear thermo–electromechanical equations based on the variational principle. The heat dissipation problem associated with piezoelectric ceramic materials was addressed by Zhou *et al.*,[16] With reasonable approximation, piezoelectric ceramics are assumed to respond linearly with little hysteresis. For PZTs, this can be up to 15%. Consider a piezoelectric ceramic block, as shown in Figure 2.3. The constitutive equation of the piezoelectric ceramic block can be expressed as follows:[17–19]

$$\{S\} = [S_E]\{T\} + [d]^T \{E\} \tag{2.1}$$

$$\{D\} = [d] \{T\} + [\varepsilon_T] \{E\} \tag{2.2}$$

where $\{S\}$ is the strain vector, $[S_E]$ is the adiabatic compliance matrix, $\{T\}$ is the stress vector, $[d]$ is the adiabatic PZT coupling matrix, $\{E\}$ is the electric field vector, $\{D\}$ is the charge density vector and $[\varepsilon_T]$ is the adiabatic dielectric constant matrix including permittivity components. Eqn (2.1) relates the applied electric field to the strain in the piezoelectric ceramic (actuator equation). On the other hand, eqn (2.2) relates the electrical charge to the stresses and strains in the piezoelectric ceramic (sensor equation). Assume that the number 1 axis is aligned in the direction in which the piezoelectric element is uniaxially stretched during its manufacturing process, and axis 3 has

two properties. While piezo polymers are stretched during manufacture, the ceramics are not. First, it is parallel to the direction of the
piezoelectric element thickness and second, the positive direction
points in the opposite direction of the electric field used in the
manufacturing process to the pole piezoelectric element (Figure 2.2).
Therefore, eqn (2.1) can be re-written as:

$$
\begin{Bmatrix} \varepsilon_1 \\ \varepsilon_2 \\ \varepsilon_3 \\ \gamma_{23} \\ \gamma_{31} \\ \gamma_{12} \end{Bmatrix}
=
\begin{bmatrix}
S_{11} & S_{12} & S_{13} & 0 & 0 & 0 \\
S_{21} & S_{22} & S_{23} & 0 & 0 & 0 \\
S_{31} & S_{32} & S_{33} & 0 & 0 & 0 \\
0 & 0 & 0 & S_{44} & 0 & 0 \\
0 & 0 & 0 & 0 & S_{55} & 0 \\
0 & 0 & 0 & 0 & 0 & S_{66}
\end{bmatrix}
\begin{Bmatrix} \sigma_1 \\ \sigma_2 \\ \sigma_3 \\ \tau_{23} \\ \tau_{31} \\ \tau_{12} \end{Bmatrix}
+
\begin{bmatrix}
0 & 0 & d_{31} \\
0 & 0 & d_{32} \\
0 & 0 & d_{33} \\
0 & d_{24} & 0 \\
d_{15} & 0 & 0 \\
0 & 0 & 0
\end{bmatrix}
\begin{Bmatrix} E_1 \\ E_2 \\ E_3 \end{Bmatrix}
$$

$$\quad\{S\} \qquad\qquad [S_E] \qquad\qquad\quad \{T\} \qquad\qquad [d]^T \qquad \{E\}$$

$$(2.3)$$

Besides the analytical models introduced in the literature, several finite
element (FE) models have also been developed. The electric field components, represented by E_k, are related to the electrostatic potential, φ:

$$E_k = -\nabla_k \varphi, \qquad k = 1,2,3 \tag{2.4}$$

or

$$
\begin{Bmatrix} S_1 \\ S_2 \\ S_3 \\ S_4 \\ S_5 \\ S_6 \end{Bmatrix}
=
\begin{bmatrix}
S_{11}^E & S_{12}^E & S_{13}^E & 0 & 0 & 0 \\
S_{21}^E & S_{22}^E & S_{23}^E & 0 & 0 & 0 \\
S_{31}^E & S_{32}^E & S_{33}^E & 0 & 0 & 0 \\
0 & 0 & 0 & S_{44}^E & 0 & 0 \\
0 & 0 & 0 & 0 & S_{55}^E & 0 \\
0 & 0 & 0 & 0 & 0 & S_{66}^E
\end{bmatrix}
\begin{Bmatrix} T_1 \\ T_2 \\ T_3 \\ T_4 \\ T_5 \\ T_6 \end{Bmatrix}
+
\begin{bmatrix}
0 & 0 & d_{31} \\
0 & 0 & d_{32} \\
0 & 0 & d_{33} \\
0 & d_{24} & 0 \\
d_{15} & 0 & 0 \\
0 & 0 & 0
\end{bmatrix}
\begin{Bmatrix} E_1 \\ E_2 \\ E_3 \end{Bmatrix}
$$

$$(2.5)$$

Note that: $s_{66}^E = 2(s_{11}^E - s_{22}^E)$.

$$
\begin{bmatrix} D_1 \\ D_2 \\ D_3 \end{bmatrix} = \begin{bmatrix} 0 & 0 & 0 & 0 & d_{15} & 0 \\ 0 & 0 & 0 & d_{24} & 0 & 0 \\ d_{31} & d_{32} & d_{33} & 0 & 0 & 0 \end{bmatrix} \begin{bmatrix} T_1 \\ T_2 \\ T_3 \\ T_4 \\ T_5 \\ T_6 \end{bmatrix} + \begin{bmatrix} \varepsilon_{11} & 0 & 0 \\ 0 & \varepsilon_{22} & 0 \\ 0 & 0 & \varepsilon_{33} \end{bmatrix} \begin{bmatrix} E_1 \\ E_2 \\ E_3 \end{bmatrix} \quad (2.6)
$$

The above constitutive equations show how the direct and inverse piezoelectric effects are governed by the nonlinear relationship between $\{S\}$ the strain vector, $[S_E]$ the adiabatic compliance matrix, $\{T\}$ the stress vector, $[d]$ the adiabatic PZT coupling matrix, $\{E\}$ the electric field vector, $\{D\}$ the charge density vector, and $[\varepsilon_T]$ the adiabatic dielectric constant matrix, in the presence of materials anisotropy. Thus, the corresponding physical quantities are described by tensors.

2.3.1 Constitutive Equations

All combinations of the various mathematical representations of the piezoelectric constitutive equations are presented below. The names for each of the forms are arbitrary; they were taken from the two dependent variables on the left-hand side of each equation. Note that the voltage and electric field variables are related *via* a gradient.

Strain–charge equations:

$$
S = S_E \cdot T + d^t \cdot E
$$
$$
D = d \cdot T + \varepsilon_T \cdot E
$$
(2.7)

Stress–charge equations:

$$
T = C_E \cdot S - e^t \cdot E
$$
$$
D = e \cdot S + \varepsilon_s \cdot E
$$
(2.8)

Strain–voltage equations:

$$
S = S_D \cdot T + g^t \cdot D
$$
$$
E = -g \cdot T + \varepsilon_T^{-1} \cdot D
$$
(2.9)

Stress–voltage equations:

$$
T = C_D \cdot S - q^t \cdot D
$$
$$
E = -q \cdot S + \varepsilon_s^{-1} \cdot D
$$
(2.10)

A list of typical PZT piezoelectric ceramic properties is presented in Table 2.1.

2.4 Applications

To date, multifunctional piezoelectric ceramics have been employed in various areas of applications for performance enhancement such as vibration/acoustic control, shape control, structural health monitoring, micro-positioning, fast valves and nozzles, transducers, gas ignitions, shock absorbers in luxury cars, active engine mounts, spacecraft jitter reduction, and sports. Several research projects have been conducted to demonstrate the application of these materials. The following is a description of some selected applications in various areas of research and industry.

2.4.1 Vibration/Acoustic Control

Vibration/acoustic control is the dominant area of applications for piezoelectric ceramics. Several vibration/acoustic applications have been investigated, including tail buffet control, wing flutter control, aircraft interior noise reduction, vibration cancellation in sports equipment such as skis, baseball bats, and snowboards. For instance, an intelligent constrained layer using a piezoelectric ceramic has been used for bending vibration control of Euler–Bernoulli beams.[20]

Undesired vibrations in typical turbine engine components such as stator and rotor blades, lubrication units, and casing reduce engine performance and life. Lucent Technologies and Pratt & Whitney have designed and applied an active vibration control system for engines to demonstrate that vibration reduction may reduce the maintenance and service costs and increase engine durability.[21] Grewal et al.[22] employed an adaptive feedforward control system to reduce the aircraft cabin noise in turboprop aircraft using piezoelectric ceramic actuators. Fripp et al.[23] also studied the application of piezoelectric ceramics in the noise control of various aircraft fuselages during operation.

Table 2.1 Typical PZT properties.

Description	Constant	Value	Unit
Volume density	ρ	7350.0	$Kg\,m^{-3}$
Elastic module	E	71.4×10^9	Pa
Piezoelectric strain constant	d_{31}	200.0×10^{-12}	$m\,V^{-1}$
Electric permittivity	ε	150.4×10^{-10}	$F\,m^{-1}$

2.4.2 Other Piezoelectric Applications

Piezo elements are used in music for acoustic instruments. They are inserted in stringed instruments such as guitars, violins, or mandolins. The dynamic deformation of the instrument (vibration of the cords) is converted into a small alternating voltage. Other applications have included skis, smart skis, a re-writable Braille alphabet for the blind, ultrasonic motors, among others.

Exercise Number 2.1 What is the mechanical quality factor, Q_m, of piezoelectric materials?

Answer to Exercise No. 2.1: The mechanical quality, Q_m, indicates how sharp the resonance is. It is determined based on the resonance of piezoelectric materials under the 3 dB bandwidth of vibrations.

Exercise Number 2.2 What is the reciprocal mechanical quality factor, Q_m, of piezoelectric materials?

Answer to Exercise Number 2.2: Consider the equivalent circuit diagram of a piezoelectric resonator at resonance. The reciprocal value of the mechanical quality factor is determined by the mechanical loss factor and the ratio of effective piezoresistance to piezoreactance.

Exercise Number 2.3 Describe the piezoelectric coefficient coupling factor vector **k**.

Answer to Exercise Number 2.3: The extent of the piezoelectric effect is measured by the coupling factor vector **k**. This vector describes the ability of a piezoelectric material to convert electrical energy into mechanical energy and *vice versa* in various modes of vibrations. The square root of the ratio of stored mechanical energy to the total energy absorbed determines the coupling factor vector **k**. The following definitions pertain to various components of the coupling factor vector **k**:

k_{33} is the coupling factor for the longitudinal oscillation, k_{31} is the coupling factor for the transverse oscillation, k_P is the coupling factor for the radial oscillation (planar) of a round disk, k_t is the coupling factor for the thickness oscillation of a plate, and k_{15} is the coupling factor of the thickness shear oscillation of a plate.

Homework Problems

Homework Problem 2.1

What is a piezo modulus?

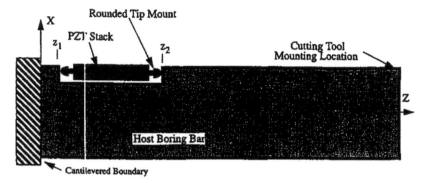

Figure 2.4 Conceptual model of a boring bar with a structurally integrated PZT stack actuator located at the bar root.

Homework Problem 2.2

Design a piezo stack of your choice composed of n actuators. Present a solution to its performance in terms of the voltage applied and displacement achieved.

Homework Problem 2.3

Assuming the host boring bar shown in Figure 2.4 has mode shapes $\phi_r(z)$ (which include the cutout), derive an expression for the modal control forces including the active and passive components of the actuator. Use the actuator force model:

$$F_a(t) = K_a(p \; d_{33} V(t) - \Delta L(t)), \tag{2.2-hw}$$

where p is the number of wafers in the stack, K_a is the actuator stiffness, d_{33} is the piezoelectric constant, $V(t)$ is the input voltage, and $\Delta L(t)$ is the axial deformation of the actuator resulting from bar bending deformations of the beam.

Homework Problem 2.4

Define the permittivity, ε.

References

1. J. Curie and P. Curie, Développement par compression de l'électricité polaire dans les cristaux hémièdres à faces inclinées, *Bull. Soc. Fr. Mineral.*, 1880, **3**, 90–93.
2. J. Curie and P. Curie, Développement, par pression, de l'électricité polaire dans les cristaux hémièdres à faces inclinées, *Comptes Rendus*, 1880, **91**, 294–295.

3. J. Curie and P. Curie, Sur l'électricité polaire dans les cristaux hémièdres à faces inclinées [On electric polarization in hemihedral crystals with inclined faces], *Comptes Rendus*, 1880, **91**, 383–386.
4. M. V. Gandhi and B. S. Thompson, *Smart Materials and Structures*, Chapman and Hall, London, 1992.
5. W. G. Cady, *Piezoelectricity*, Dover Publications, Inc., New York, 1964.
6. W. G. Cady, *Piezoelectricity: An Introduction to the Theory and Applications of Electromechanical Phenomena In Crystals*, McGraw-Hill, New York, 1946.
7. B. Z. Janos and N. W. Hagood, Internal report # AMSL 00-4, Active Materials and Structures Laboratory, MIT, 2000.
8. Q. M. Zhang, V. Bharti and X. Zhao, Giant Electrostriction and Relaxor Ferroelectric Behavior in Electron-Irradiated Poly(vinylidene fluoride-trifluoroethylene) Copolymer, *Science*, 1998, **280**, 2101–2104.
9. ANSI-IEEE 176, Standards on Piezoelectricity, Institution of Electrical and Electronics Engineers (IEEE), 1987.
10. V. Giurgiutiu, Review of Smart-Materials Actuation Solutions for Aeroelastic and Vibration Control, *J. Intell. Mater. Syst. Struct.*, 2000, **11**, 525–544.
11. S. Qi, R. Shuttleworth, S. Olutunde Oyadiji and J. Wright, Design of a multi-resonant beam for broadband piezoelectric energy harvesting, *Smart Mater. Struct.*, 2010, **19**, 1–10.
12. S. Rafique and P. Bonello, Experimental validation of a distributed parameter piezoelectric bimorph cantilever energy harvester, *Smart Mater. Struct.*, 2010, **19**(15pp), 094008.
13. A. Yousefi-Koma and G. Vukovich, Dynamic Modelling of Piezo transducers for Flexible Smart Structures, *Can. Aeronaut. Space J.*, 1999, **45**(4), 379–389.
14. K. H. Chan and N. W. Hagood, *Proceedings of the SPIE Conference on Smart Structures and Materials: Smart Structures and Intelligent Systems*, Orlando, Florida, 1994, pp. 195–205.
15. H. S. Tzou, Y. Bao and R. Ye, *Proceedings of the SPIE Conference on Smart Structures and Materials: Smart Structures and Intelligent Systems*, Orlando, Florida, 1994, pp. 206–214.
16. S. Zhou, C. Liang and C. A. Rogers, *Adaptive Structures and Composite materials: Analysis and Application, ASME International Mechanical Engineering Congress and Exposition, Chicago, Illinois, November 6–11*, 1994, pp. 183–191.
17. D. A. Berlincourt, D. R. Curran and H. Jaffe, *Physical Acoustics Principles and Methods*, ed. W. P. Mason, Academic Press, New York, 1964, pp. 169–270.
18. B. Jaffe, R. Cook and H. Jaffe, *Piezoelectric Ceramics*, Academic Press, New York, NY, 1971.
19. J. Zelenka, *Piezoelectric Resonators and Their Applications*, Elsevier Science Publishing Co., Inc., New York, 1986.
20. I. Y. Shen, Stability and Controllability of Euler–Bernoulli Beams With Intelligent Constrained Layer Treatments, *J. Vib. Acoust.*, 1996, **118**(1), 70–77.
21. E. D. Finn, Lowering the Volume on Helicopters, *Aerospace America*, 1997, 24–25.
22. A. Grewal, D. G. Zimcik and B. Leigh, Feedforward Piezoelectric Structural Control: An Application to Aircraft Cabin Noise Reduction, *J. Aircr.*, 2001, **38**(1), 164–173.
23. M. Fripp, D. O'Sullivan, S. Hall, N. W. Hagood and K. Lilienkamp, Applications of piezoelectric ceramics in noise control of various aircraft fuselages during operation, *Proceedings of SPIE Smart Structures and Materials*, vol. 3041, 1997.

3 Review of Piezoresistive Materials as Smart Sensors

Mohsen Shahinpoor

Mechanical Engineering Dept., University of Maine, USA
Email: shah@maine.edu

3.1 The Piezoresistivity Effect

Piezoresistivity is defined as a property of certain materials, such as metals and semiconductors, for which the materials electrical resistance changes purely as a result of mechanical pressure, stress, force, acceleration, strain, and stress. It is a physical property of certain materials with applications as smart sensors, accelerometers, tactile sensors, strain gauges and flows meters, and similar devices and microdevices that reply on the conversion of a mechanical signal (stress, strain) to an electrical signal (voltage, current). The unit of piezoresistivity is the ohm-meter or symbolically Ω–m.

Metals and semiconducting materials exhibit such a property. Metals do not exhibit piezoresistivity as they do not have a bandgap. The resistance of strained metal samples changes due to dimensional changes – this is not considered as true piezoresistivity as the resistivity of the material itself is not changing. The piezoresistive effect in semiconductors is generally much larger than the geometrical effect. This effect can be observed in semiconductors such as germanium, amorphous silicon, silicon carbide, and polycrystalline silicon, among others. Thus, semiconductor strain gauges with high sensitivity can be designed, built, operated, and utilized in various

Fundamentals of Smart Materials
Edited by Mohsen Shahinpoor
© The Royal Society of Chemistry 2020
Published by the Royal Society of Chemistry, www.rsc.org

smart sensor applications. They can also serve as smart microelectromechanical (MEMs) or nanoelectromechanical (NEMs) devices and
systems.

In this sense, piezoresistivity describes the changing resistivity of a
semiconductor due to applied mechanical stress, strain, or force.
Note that, compared to the piezoelectric effect, the piezoresistive effect does not produce any charge or electric potential upon strain or
stress and does not show any deformation under an applied electric
field, only producing a change in resistance upon strain, stress or
acceleration. It was Lord Kelvin in 1856 who discovered the change
in resistance in metals such as iron and copper due to an applied
mechanical load. His experiments were motivated by challenges
caused by conductivity changes in telegraph wires. In 1954, Smith
first reported the large piezoresistive effect in silicon and germanium.[1] If an applied stress or strain field changes the interatomic
spacing in a semiconductor, then the bandgap in the semiconductor
changes. This provides the energy required to convert a valence
electron bound to an atom to a conduction electron. Conduction
electrons are free to move within the crystal lattice and serve as charge
carriers to conduct electric current, making it possible for electrons to
enter into the conduction band and thus change the conductivity of
the semiconductor. Because semiconductor strain gauges are generally more sensitive to environmental conditions such as humidity and
temperature, it is more difficult to handle them compared to conductor (metal) strain gauges.

Note that piezoresistivity, ρ_σ, due to applied stress, σ, can be
expressed using eqn (3.1):

$$\rho_\sigma = (\partial\rho/\rho)/\varepsilon \qquad (3.1)$$

where $\partial\rho$ is the change in resistivity, ρ is the initial resistivity, and ε is
the strain. The changes in metallic resistance are due to the changes
in the geometry of the metal resulting from the applied mechanical
stress, σ. Note that semiconductors such as silicon or germanium create
much more sensitive strain gauges. The resistance, R, can be calculated
using the simple resistance equation derived from the Ohm's law of a
voltage drop across a resistor R carrying a current I, $v = RI$:

$$R = \rho l/A \qquad (3.2)$$

where l is the conductor length in the meter and A is the cross-
sectional area of the wire in m^2. Piezoresistivity is numerically described by what is called a gauge factor (GF) or sometimes denoted by

$\gamma = \mathrm{GF}$, which is defined as the fractional change in the resistance R per unit strain ε:

$$\mathrm{GF} = \gamma = \Delta R / R\varepsilon \qquad (3.3)$$

where GF is the fractional change, R is the nominal electrical resistance, and ε is the strain. Note that GF is a dimensionless number depending on the crystallographic orientation of the crystalline material. It is related to the Young's modulus of elasticity E of the piezoresistive material by the following expression (Hooke's law):

$$E = \sigma / \varepsilon \qquad (3.4)$$

3.2 Piezoresistive Strain/Stress Sensor Configuration

The variations in the electrical resistance of the piezoresistors are not sensed directly but through a traditional Wheatstone bridge. In this case, the piezoresistors are wired together with an electrical circuit configuration known as the Wheatstone bridge (Figure 3.1).

The bridge has a constant input voltage, V_{EX}, and produces an output voltage, V_0, that is related to the value of the resistance in the bridge. Note that replacing R_4 in the Wheatstone bridge circuit in Figure 3.1 with an active piezoresistive strain gauge causes any changes in the gauge resistance to unbalance the bridge and produce a nonzero output voltage, V_0. The strain-induced variation in resistance, ΔR, can be expressed as $\Delta R = R_G \cdot \mathrm{GF} \cdot \varepsilon$, where GF is the previously defined gauge factor, and R_G is the nominal resistance of the strain gauge. Assuming $R_1 = R_2$ and $R_3 = R_G$, the bridge equation can be rewritten to express V_0 / V_S as a function of strain, as shown in

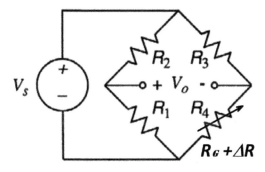

Figure 3.1 Piezoresistive quarter Wheatstone bridge circuit.

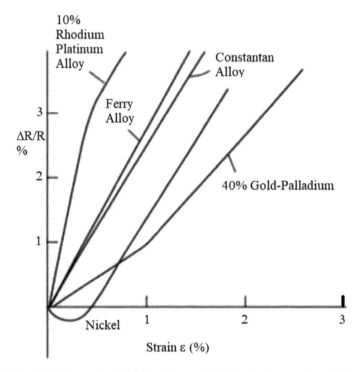

Figure 3.2 Variations in ($\Delta R/R$) with applied strain for various thin metallic films.
Reprinted from ref. 2, with the permission of AIP Publishing.

eqn (3.5). Applying Kirchhoff's Law to the quarter Wheatstone bridge we obtain:

$$(V_0/V_s) = (\text{GF·}\varepsilon/4)\{1/(1 + \text{GF·}\varepsilon/2)\}^{-1} \tag{3.5}$$

The presence of the quotient $\{1/(1 + \text{GF·}\varepsilon/2)\}$ term clearly shows the nonlinearity of the quarter-bridge output concerning the strain ε. Thus, stresses and strains can be measured by calibration of the voltage output, V_0, according to the applied strain, ε, obtained by the application of the piezoresistive strain gauges.

According to the experimental results obtained by Parker and Krinsky in 1963,[2] the electrical resistance variations of the thin metallic films are presented in Figure 3.2.

3.3 Piezoresistive Strain Sensors

Piezoresistive strain sensors measure the strain by the observed changes in their conductivity or resistance, which correlate to the applied strain or stress. The fundamental principle is essentially the symmetry breaking of the semiconductor crystalline lattice due to the applied strain or stress. The symmetry breaking strain of the semiconductor crystal lattice distorts the energy bands, and splits the underlying energy levels, changing the carrier scattering rates, which ultimately changes semiconductor resistance. Thus, piezoresistive strain sensors are designed to measure the applied strain or stress by producing a proportional change in their resistance.

For a piezoresistive transducer, one can plot the measured change in resistance against the applied strain to figure out the input strain. In this case, the upper limit of the measurable strain is normally defined as the maximum strain beyond which nonlinear strains manifest themselves.

3.4 Physical Causes of Piezoresistivity

Change in relative dimensions, as the resistance is related to length and cross-sectional area of the sample, produces piezoresistivity. The relevant equations are:

$$R = (\rho l/A) \text{ or } dR = (\rho/A)\, dl + (l/A)\, d\rho - (\rho l/A^2)\, dA \qquad \text{(3.3a and b)}$$

$$(dR/R) = (dl/l) + (d\rho/\rho) - (dA/A) \qquad \text{(3.4)}$$

3.4.1 Basic Formula for Describing Piezoresistivity

Some of the basic governing equations of piezoresistivity are written in terms of the gauge factor, GF, which is the ratio of relative change in resistance due to strain. Thus,

$$(\Delta R/R) = GF\, (\Delta l/l) \qquad \text{(3.5)}$$

$$GF = (\Delta R/R)/(\Delta l/l) = (\Delta R/\varepsilon R) \qquad \text{(3.6)}$$

$$\text{Stress} = \sigma = \varepsilon E \qquad \text{(3.7)}$$

where, E is the Young's modulus of elasticity of the piezoresistive material. These equations are used to design piezoresistive systems.

3.4.2 Relationship Between the Change in Resistance {ΔR} and Stresses {σ}

Consider the change in resistance in an infinitesimal differential element of a piezoresistive crystal element such that on each side of the microcube of a piezoresistive material there are 6 components of normal and shear stresses, given by eqn (3.8):

$$\boldsymbol{\sigma} = \{\sigma_{xx} \ \sigma_{yy} \ \sigma_{zz} \ \sigma_{xy} \ \sigma_{xz} \ \sigma_{yz}\}^{\mathrm{T}} \tag{3.8}$$

such that the change in resistance on each surface of the microelement is a vector, $\Delta\mathbf{R}$, such that

$$\Delta\mathbf{R} = [\zeta]\{\boldsymbol{\sigma}\} \tag{3.9}$$

where $\{\Delta\mathbf{R}\} = \{\Delta R_{xx} \ \Delta R_{yy} \ \Delta R_{zz} \ \Delta R_{xy} \ \Delta R_{xz} \ \Delta R_{yz}\}^{\mathrm{T}}$ is the vector showing the changes in resistances in various directions in a small cubic piezoresistive crystal element with corresponding stress components $\sigma_{xx} \ \sigma_{yy} \ \sigma_{zz} \ \sigma_{xy} \ \sigma_{xz} \ \sigma_{yz}$, and $[\zeta]$ represents the matrix of the piezoresistive coefficients. Expanding eqn (3.9) yields the following constitutive equations relating the changes in resistance due to applied stresses in a piezoresistive material, which result in the following:

$$\Delta R_{xx} = \zeta_{11}\sigma_{xx} + \zeta_{12}(\sigma_{yy} + \sigma_{zz}), \Delta R_{yy} = \zeta_{11}\sigma_{yy} + \zeta_{12}(\sigma_{xx} + \sigma_{zz}),$$

$$\Delta R_{zz} = \zeta_{11}\sigma_{zz} + \zeta_{12}(\sigma_{xx} + \sigma_{yy}) \tag{3.10a-f}$$

$$\Delta R_{xy} = \zeta_{44}\sigma_{xy}, \ \Delta R_{xz} = \zeta_{44}\sigma_{xz}, \ \Delta R_{yz} = \zeta_{44}\sigma_{yz}$$

and thus only 3 piezoresistive coefficients are required; ζ_{11} and ζ_{12} associated with normal stresses and ζ_{44} associated with shearing stresses. In matrix form, eqn (3.10) is represented by:

$$\begin{Bmatrix} \Delta R_1 \\ \Delta R_2 \\ \Delta R_3 \\ \Delta R_4 \\ \Delta R_5 \\ \Delta R_6 \end{Bmatrix} = \begin{bmatrix} \zeta_{11} & \zeta_{12} & \zeta_{13} & 0 & 0 & 0 \\ \zeta_{21} & \zeta_{22} & \zeta_{23} & 0 & 0 & 0 \\ \zeta_{31} & \zeta_{32} & \zeta_{33} & 0 & 0 & 0 \\ 0 & 0 & 0 & \zeta_{44} & 0 & 0 \\ 0 & 0 & 0 & 0 & \zeta_{55} & 0 \\ 0 & 0 & 0 & 0 & 0 & \zeta_{66} \end{bmatrix} \begin{Bmatrix} \sigma_{xx} \\ \sigma_{yy} \\ \sigma_{zz} \\ \sigma_{yz} \\ \sigma_{zx} \\ \sigma_{xy} \end{Bmatrix} \tag{3.11}$$

3.4.3   Why Use Semiconductor Strain Gauge?

There are certain advantages in using semiconductors as strain gauges due to the fact that they possess higher GFs than metallic alloy strain gauges, they can be easily fabricated with controlled performance specifications using precise ion implantation and diffusion, and they can be easily integrated with silicon, a material used for sensors and signal processing.

3.5 Merit of Piezoresistive Sensors *vs.* Capacitive Sensors

Capacitive sensing is perhaps the most dominant position-sensing technique for microfabricated sensors. However, there are several limitations to capacitive sensors. The detection of position is constrained to the small vertical movement (parallel plate) and horizontal movement (transverse or lateral comb drives). The area of overlapped electrodes must be reasonably large (as a rule of thumb, tens of mm^2). If the overlap area is small and the vertical displacement is large, capacitive sensors are not suitable.

Charles Smith observed the piezoresistive effect in germanium and silicon in 1954. His test samples included As-doped n-type germanium, Ga-doped p-type germanium, and both n and p types of silicon crystals. He reported that one of the shear coefficients for each sample was too large to be explained by mechanisms previously determined in metals displaying piezoresistivity. Previous research had indicated that piezoresistance arose from changes in either the mobility or number of charge carriers in the material. Smith noted that mobility changes were too small to influence the shear coefficients, and the change in several carriers should not have had any effect on the shear coefficients.[1] Smith also observed some dependence of the piezoresistive coefficients upon the degree of doping.

In p-silicon with a resistance of 7.8 $\Omega\,cm^{-1}$, Smith reported a piezoresistive tensor with the components $\zeta_{11} = \zeta_{22} = \zeta_{33} = 6.6 \times 10^{-12}\,cm^2\,dyne^{-1}$, $\zeta_{12} = \zeta_{23} = \zeta_{13} = -1.1 \times 10^{-12}\,cm^2\,dyne^{-1}$, and $\zeta_{44} = 138.1 \times 10^{-12}\,cm^2\,dyne^{-1}$.[1] In his 1954 paper, Smith noted that the model agrees with the results for n-silicon, but not for p-germanium. Smith suggested an additional mechanism was needed to explain the behavior of p-germanium. The investigation of piezoresistive behavior in p-type materials has required recent computational advances.[3] Matsuda *et al.* (1993) reported on nonlinearity in the piezoresistive effect exhibited by p and n-type silicon crystals.[4] Prior work had been done on the nonlinearity of p-silicon because most piezoresistive sensors are constructed from p-silicon. It has been observed that the complexity of the valence bands in p-silicon makes interpreting the piezoresistive effects hard, while in n-silicon the first-order piezoresistive coefficients arose from carrier transfer and were much easier to determine.[4] Presumably, the dominance of p-silicon in sensor applications is due to fabrication concerns outside the concern of Matsuda's work. Berg *et al.*[5] tested the

piezoresistance of graphite fibers subjected to tensile loads or torsion.[5] Their experiments indicated a relationship between the piezoresistance of the fiber and its modulus of elasticity.[3,6] Whittaker Type
II fibers, with a modulus of 4.8×10^7 *psi*, exhibited an increase in resistance, while Celanese fibers, with a modulus of 11.4×10^7 *psi* exhibited a decrease in resistance. Their observation of the Whittaker
fibers' behavior was consistent with observations on Morganite type
fibers. The behavior of the Celanese fibers was suggested to be due to
differing defect structures between the Celanese and Whittaker fibers,
although insufficient information on their structures was available at
the time for verification. The torsion tests indicated an increase in
resistance under torsion, but the process was not reversible.[3] Berg *et al.*
commented that piezoresistance in graphite fibers might allow for the
construction of fiber composites that act as their strain gauges.[5]

In 2006, He and Yang[7] reported giant piezoresistance in silicon
nanowires, exceeding known responses by two orders of magnitude.
A literature review by Rowe (2014) maintains that the giant piezoresistivity
effect in nanowires appears to be reproducible in gated nanowires.[8]

As previously noted, most piezoresistors used commercially are of
the p-type. Doping options include diffusion, ion implantation, and
epitaxy, with ion implantation being the most common technique
used due to superior control.[3] While the earliest experiments conducted by Smith involved fairly low doping amounts, this is advantageous for some sensors, as it reduces the sensor's sensitivity to
temperature variations, as well as reducing nonlinearities in the
sensor's response to temperature and strain.[3]

Pressure sensors utilizing piezoresistors have been available since
the late 1950s. Pressure sensors developed throughout the 1960s and
1970s, and from the 1980s onward fabrication techniques related to
MEMS applications has resulted in significant improvements in size
and performance.[3,6]

3.6 Piezoresistivity Components

Components of $[\zeta]$ representing the matrix of the piezoresistive
coefficients are given below in connection with piezoresistivity components for single-crystal silicon under certain doping values:

n-type ζ_{11} with a resistivity of 11.7 Ωcm $= -103.2 \times 10^{-11}$ Pa^{-1}
p-type ζ_{11} with a resistivity of 7.8 Ωcm $= 6.6 \times 10^{-11}$ Pa^{-1}
n-type ζ_{12} with a resistivity of 11.7 Ωcm $= 53.4 \times 10^{-11}$ Pa^{-1}

p-type ζ_{12} with a resistivity of 7.8 Ωcm $= -1.1 \times 10^{-11}$ Pa^{-1}
n-type ζ_{44} with a resistivity of 11.7 Ωcm $= -13.6 \times 10^{-11}$ Pa^{-1}
p-type ζ_{44} with a resistivity of 7.8 Ωcm $= 138.1 \times 10^{-11}$ Pa^{-1}

Exercise 3.1: Estimate the change in resistance in silicon piezo-resistors connected to a pressure vessel of maximum pressure of p such that the maximum stress in the photoresistor is $\sigma_{max} = 186.81$ MPa
   Solution to Exercise 3.1: Let: $\sigma_{max} = 186.8$ MPa, or $= 186.8 \times 10^6$ Pa $(\mathrm{N\,m^{-2}})$

Let the two piezoresistive coefficients be: $\zeta_{11} = \zeta_{12} = 0.02\ \zeta_{44}$
But $\zeta_{44} = 138.1 \times 10^{-11}$ Pa^{-1} from the above section and thus,
$\Delta R / R = \zeta_{11}\ \sigma_{max} + \zeta_{12}\ \sigma_{max} = 2 \times 0.02\ (138.1 \times 10^{-11})\ (186.8 \times 10^6) =$ 0.01032 Ω / Ω

3.7 Methods for Compensating for the Temperature Effect

Doped silicon strain sensors are also sensitive to temperature. To study the changes in the strain field due to temperature variation, one should measure and record both variations.

- Common technique: use a reference resistor that is subject to the same temperature but not the strain. The difference in signals between these two piezoresistive sensors gives the overall effect due to strain.
- Second technique: Wheatstone bridge

The Wheatstone bridge circuit transforms resistance change to voltage change:

$$V_{out} = \left(\frac{R_2}{R_1 + R_2} - \frac{R_4}{R_3 + R_4} \right) V_{in} \quad \text{or} \quad V_{out} = \left(\frac{-\Delta R / 2}{2R + \Delta R} \right) V_{in}$$

which is temperature insensitive

$$V_{out} = \left(\frac{R}{R + (R + \Delta R)} - \frac{R}{2R} \right) V_{in} = \left(\frac{R}{2R + \Delta R} - \frac{1}{2} \right) V_{in}$$

$$= \left(\frac{R}{2R + \Delta R} - \frac{\left(R + \frac{\Delta R}{2} \right)}{2 \left(R + \frac{\Delta R}{2} \right)} \right) V_{in}$$

Exercise 3.2: A p-type silicon piezoresistor with certain doping develops a 30% per °C of its piezoresistivity at 20 °C. Then, at an operating temperature of 120 °C, assuming an initial temperature of 20 °C, it would lose $(120 - 20) \times 0.3\% = 30\%$ of the value of the piezoresistivity coefficient. For example: Estimate the change in resistance in silicon piezoresistors attached to the square diaphragm of a pressure sensor under a maximum pressure of $\sigma_{max} = 200$ MPa. Determine the relative loss of resistance $\Delta R/R$ if the total piezoresistive coefficient is $= 5 \times 10^{-11}$ Pa^{-1}.

The solution to Exercise 3.2: The relative loss of resistance is $\Delta R/R = \zeta \ \sigma_{max}$. Thus, in this case we obtain: $\Delta R/R = \zeta \ \sigma_{max} = 5 \times 10^{-11}$ Pa$^{-1} \times 200 \times 10^6$ Pa $= 0.01$ Ω/Ω

Summary

In summary, it should be mentioned that the long history of piezoresistance has seen comparatively recent developments in commercial applications that use semiconductors, in particular silicon p- and n-type crystals. The giant piezoresistance effects still require further investigation to be developed for sensor applications. Conventional piezoresistors are now widely available and used for pressure and force sensing applications. The sensitivity they provide make them the superior choice for sensor applications, and continued research will provide a better understanding of their response behavior.

Homework Problems

Homework Problem 3.1

A piezoresistor is embedded on the top surface of a silicon cantilever near the cantilevered base. The cantilever points in the <110> direction. The piezo resistor is a doped p-type and has a resistivity of 7.8 Ωcm. Find the longitudinal gauge factor of the material.

Homework Problem 3.2

A piezoresistor is embedded on the top surface of a silicon cantilever near the cantilevered base. The cantilever points in the <110> direction. The piezoresistor is an n-type doped with a resistivity of 7.8 Ωcm. Find the longitudinal gauge factor of the material.

References

1. C. S. Smith, Piezoresistance effect in germanium and silicon, *Phys. Rev.*, 1954, **94**(1), 42–49.
2. R. L. Parker and A. Krinsky, Electrical resistance-strain characteristics of thin evaporated metal films, *J. Appl. Phys.*, 1963, **34**(9), 2700.
3. A. A. Barlian, W.-T. Park, J. R. Mallon, A. J. Rastegar and B. L. Pruitt, Review: Semiconductor Piezoresistance for Microsystems, *Proc. IEEE Institut. Elec. Electron. Engin.*, 2009, **97**(3), 513–552.
4. K. Matsuda, K. Suzuki, K. Yamamura and Y. Kanda, Nonlinear piezoresistance effects in silicon, *J. Appl. Phys.*, 1993, **59**(4), 1838–1847.
5. C. A. Berg, H. Cumpston and A. Rinsky, Piezoresistance of graphite fibers, *Text. Res. J.*, 1972, **42**(8), 486–489.
6. J. Bryzek, Impact of MEMS technology on society, *Sens. Actuators, A*, 1996, **56**(1–2), 1–9.
7. R. He and P. Yang, Giant piezoresistance effect in silicon nanowires, *Nat. Nanotechnol.*, 2006, **1**, 42–46.
8. A. C. H. Rowe, Piezoresistance in silicon and its nanostructures, *J. Mater. Res.*, 2014, **29**(6), 731.

4 Review of Electrostrictive Materials

Mohsen Shahinpoor

Mechanical Engineering Dept., University of Maine, USA
Email: shah@maine.edu

4.1 Introduction

Electrostriction is defined as the electromechanical coupling in all electrical-nonconductor (dielectric, insulator) materials. Under the application of an electric field, these materials show deformation, strain, and stress. Generally speaking, all electrostrictive materials exhibit second-order nonlinear coupling between the elastic strains or stresses and the dielectric terms such that the strain tensor is given by a nonlinear product of the vectors of the electric fields. For a single uniaxial strain (deformation), the induced strain (deformation) is directly proportional to the square of the applied electric field (voltage). Electrostrictive materials in the form of polymers have been the subject of much interest and research in recent years, and much of the focus has been on actuator configurations and how to enhance electromechanical activities.

When an electric field is applied to a dielectric material, the differential elements and domains of the material get polarized and opposite sides of these domains become differently charged and attract each other, reducing material thickness in the direction of the applied field (and increased thickness in the orthogonal directions due to the Poisson's ratio). The resulting strain tensor S_{ij} ($i,j = 1, 2\ 3$)

Fundamentals of Smart Materials
Edited by Mohsen Shahinpoor
© The Royal Society of Chemistry 2020
Published by the Royal Society of Chemistry, www.rsc.org

is proportional to the product of the polarization vectors P_k. For simple one-dimensional domains, the strain is proportional to the square of the applied electric field.

For 3D modeling of electrostrictions, the constitutional relationships are generally highly nonlinear and coupled. In fact, the electrostrictive strain tensor, ε_{ij}, associated with the induced polarization components P_k and P_l is given by $S_{ij} = Q_{ijkl}P_kP_l$, where Einstein's summation convention is applied for all repeated indices and Q_{ijkl} are the electrostrictive coefficients, which have values in the range of 10^{-3} m^{4}C^{-2} in relaxor ferroelectrics to 10^3 m^{4}C^{-2} in dielectric polymers.[1]

Electrostrictive actuators operate similarly to piezoceramic actuators such as lead zirconate titanate (PZT), but in a nonlinear manner. Aside from PZT as the most widely used piezoelectric actuator and commercially available sensor system, lead magnesium niobate [Pb (Mg$_{1/3}$Nb$_{2/3}$) O$_3$], also known as PMN, is commercially available and probably the most studied electrostrictive material.[2,3]

Another electrostrictive material similar to PMN is lead titanate [PbTiO$_3$], also known as PT as well as the combination of the two, known as PMN–PT. PMN does not need to be polarized. Unlike piezoceramics such as PZT, electrostrictive ceramics do not need to be polarized. PMN is also a relaxor ferroelectric. Relaxor ferroelectrics have a very high dielectric constant (D > 20 000). Strain varies nonlinearly (quadratically) with the electric field for an electrostrictive material such as PMN, rather than linearly as is the case for piezoelectric materials. Electrostrictive actuators must be operated above the Curie temperature, which is typically very low when compared to those of piezoceramics such as PZTs.

Figure 4.1 depicts the fundamental difference between piezoceramic actuation and electrostrictive actuation. The relationship between the drive voltage, V, and the displacement, ΔL, is quadratic, which means that PMN actuators are intrinsically nonlinear, in contrast to PZT actuators.

According to ref. 4 and 5, PMN actuators have an electrical capacitance several times higher than those of piezo actuators and hence require higher drive currents for dynamic applications. However, electrostrictive actuators exhibit less hysteresis (approximately 3% or so compared to 12–15% for PZT).

Note that PZT materials have greater temperature stability than electrostrictive materials in the sense that as the temperature increases, possible movement is reduced/diminished, while at low temperatures, the movement is the greatest but the hysteresis

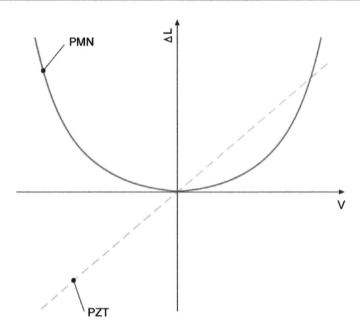

Figure 4.1 Comparison of voltage–displacement relationships in PMN and PZT materials.

increases. PMN actuators are thus best for applications with little or no temperature variations.

According to TRS Technologies (http://www.trstechnologies.com),[2] for their PMN-15, electrostriction can produce a strain of 0.1% at a field strength of 2 million volts per meter (2 MV m^{-1}). The effect is quadratic at low field strengths (up to 0.3 MV m^{-1}) and nearly linear above 0.3 MV m^{-1}, up to a maximum field strength of 4 MV m^{-1}.

Exercise No. 4.1 How do piezoelectric materials differ from electrostrictive materials in terms of capacitance and creep?

Answer to exercise 4.1 PMN actuators have an electrical capacitance several times higher than those piezoelectric actuators. However, electrostrictive actuators exhibit less hysteresis (in the order of 3% or so compared to 15% or so), than piezo actuators.

4.2 Constitutive Equations and Electrostrictive Properties

As discussed in the introduction, electrostrictive materials such as PMN or PMN–PT, also known as ferroelectric relaxors or ferro-relaxors, exhibit a property called electrostriction, a property of all dielectric materials. Electrostriction causes the material to mechanically deform under the application of an electric field, similar to piezoelectric materials. Under the electric field, the randomly oriented domains align along the same direction of the applied field, causing the deformation. Both electrostrictive and piezoelectric materials convert electrical energy into mechanical energy by mechanically deforming under an applied field or voltage. Their response, however, differs both in magnitude and direction. Piezoelectric materials have a linear response to an applied voltage, while electrostrictive materials have a displacement proportional to the square of the voltage. This is shown by the constitutive eqn (4.1) and (4.2) for piezoelectrics and eqn (4.3) and (4.4) for electrostrictive materials.

Electrostriction is generally defined as the quadratic coupling between the strain (S_{ij}) and polarization (P_m):[6,7]

$$E_m = \varepsilon^* T_{mn}\, P_n + 2Q_{klmn}\, T_{kl}\, P_n \tag{4.1}$$

$$S_{ij} = {_s}P_{ijkl}\, T_{kl} + d_{kij}\, E_k + Q_{ijmn}\, P_m\, P_n \tag{4.2}$$

where ${_s}P_{ijkl}$ is the elastic compliance, Q_{ijkl} is the polarization-related electrostriction coefficient, $\varepsilon^* T_{mn}$ is the inverse of the linear dielectric permittivity, T_{kl} is the stress and E_m the electric field. A linear relationship is assumed between the polarization and the electric field. The strain, S_{ij}, and the electric flux density, D_i, are expressed as independent variables of the electric field intensity E_k, E_l, and stress T_{kl} by the constitutive relationships according to the equations reported in ref. 8 and 9 where S^E_{ijkl} is the elastic compliance under constant electric field, M_{ijkl} is the electric field-related electrostriction coefficient, and ε^T_{ik} is the linear dielectric permittivity.

$$S = S^E \times T + d \times E \tag{4.3}$$

$$D = \varepsilon^T \times E + T \times E \tag{4.4}$$

$$S_{ij} = {_s}E_{ijkl}\, T_{kl} + d_{kij}\, E_k + m_{klij}\, E_k\, E_l \tag{4.5}$$

$$D_m = d_{mkl}\, T_{kl} + {_\varepsilon}T_{mn}\, E_n + 2m_{mnij}\, E_n\, T_{ij} \tag{4.6}$$

$$E_m = \varepsilon^* T_{mn}\, P_n + 2Q_{klmn}\, T_{kl}\, P_n \tag{4.7}$$

$$S_{ij} = {_s}P_{ijkl}\, T_{kl} + d_{kij}\, E_k + Q_{ijmn}\, P_m\, P_n \tag{4.8}$$

$$S = S^E \times T + m \times E^2 \tag{4.9}$$

$$D = \varepsilon^T \times E + m \times T \times E \tag{4.10}$$

where S is the strain tensor, S^E the compliance tensor, T the strain tensor, d the piezoelectric coefficients, E the electric field, D is the electric displacement, ε^T the dielectric permittivity tensor at constant stress, and m the electrostrictive coefficients. Eqn (4.3) shows that the strain of an electrostrictive material in the same direction as the electric field is positive, regardless of the polarity of the field, and will not contract from its original size. On the other hand, for a piezoelectric material, a negative voltage, or a voltage in the opposite direction of the poles, induces a contraction. This is shown graphically in Figure 4.1.

Electrostrictive materials have less hysteresis than piezoelectrics, meaning that there is lower energy loss as the voltage is applied and removed, making them better suited to very high frequencies. The dielectric permittivity of electrostrictive materials is larger than the permittivity of piezoelectric materials (as much as a factor of 10 or more than for PZT), which result in strains of about an order of magnitude larger than in piezoelectrics.

Not all of the physical properties of electrostrictive materials are better than those of piezoelectric materials. As shown in Figure 4.1, the displacement of electrostrictive materials is not linear, making it more difficult to characterize. Electrostrictive materials are highly temperature dependent and have a narrow temperature range (typically 10–20 °C) over which they can operate usefully, making them less suitable for some field applications.

Exercise 4.2 How do PZT and PMN compare in terms of applied positive or negative electric fields?

Answer to Exercise 4.2 Note that eqn (4.3) shows that the strain of an electrostrictive material in the same direction as the electric field is positive, regardless of the polarity of the field, and will not contract from its original size (Figure 4.1). On the other hand, for a piezoelectric material, a negative voltage, or a voltage in the opposite direction of the poles will induce a contraction.

4.3   PMN Impedance Mismatch

PMN is the most common of the electrostrictive materials and was used in early sonar applications. Doping it with PT makes PMN–PT, which is now widely used in ultrasonics for medical use and non-destructive testing (NDT).

Due to a 20:1 impedance mismatch, the high acoustical impedance of single-crystalline PMN–PT makes ultrasonic imaging of human tissue very difficult. One method used to alter the PMN–PT impedance mismatch is to dice a single-crystalline PMN–PT wafer with a dicing saw and filling the cut areas with an epoxy matrix, resulting in a 1–3 PMN–PT composite, which is 1–3 composites of PMN–PT single crystals in a soft epoxy matrix.

4.4 PMN Suppliers

TRS Technologies in Pennsylvania, US, has three PMN–PT materials available.[2] The materials operate over specific temperature ranges. The responses of PMN-15, one of TRS's available materials, is shown in Figure 4.2 over different temperatures, where it can be seen that the best response is at 2 °C. Meanwhile, in the United Kingdom, Ecertec has electrostrictive powders and ceramics for sale.[3]

Exercise 4.3 How do temperature variations affect the performance of electrostrictive materials?

Answer to Exercise 4.3 According to Figure 4.2, the higher the temperature (below the glass transition), the higher the strain under the same stress.

4.5 Electrostrictive Materials Compared to Piezoelectric Materials

It is illuminating to note that piezoelectric actuators are important in many micro-positioning systems. However, because of their inherently

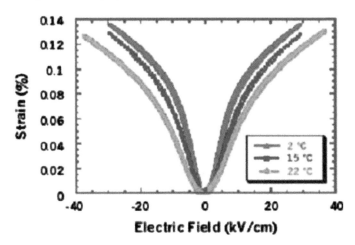

Figure 4.2 Temperature responses of PMN-15.
Reproduced from ref. 2 with permission from TRS Technologies, Copyright 2018.

high hysteresis and creep, they lack repeatability. Thus, electrostrictive actuators have been developed that provide open-loop positioning repeatability improvements of an order of magnitude or more.

These actuators offer minimum incremental motion of as fine as 0.02 μm. The performance improvement of electrostrictive actuators is due to the characteristics of PMN, a ferroelectric material with superior properties for use in motion control applications.

Electrostrictive materials are still ferroelectric crystals, which deform in proportion to an applied voltage. We note that electro-strictive actuators use a PMN crystal stack. The PMN stack has a multi-layered configuration with very thin layers (125–250 μm) that are diffusion bonded during the manufacturing process. The total positive displacement of the PMN stack is the superposition of the strain from the individual layers. The change in length for PMN materials is proportional to the square of the voltage applied and on the same order as for that of PZT. Unlike those of PZT, PMN ceramics are not poled, therefore, alternating between a negative or positive voltage results in elongation in the direction of the applied field, regardless of polarity. The PMN is an inherently more stable material since it is not poled like piezoelectrics and its creep is reduced to just 3%, while this climbs up to 15% in PZT devices.[9,10] As far as actuation under an electric field is concerned, electrostrictive materials such as PMN lead to a larger displacement under the same electric field compared to PZTs.[2,4,9–11]

PMN materials have better characteristics than conventional PZT compounds. Piezoelectric devices exhibit a hysteresis of 12–15%, while the hysteresis of PMN is 2%. Figure 4.3 depicts the difference between the hysteresis in piezoelectric and electrostrictive materials.[9]

Two properties of ferroelectric materials contribute to their thermal stability. The PMN coefficient of thermal expansion is twice that of PZT at an expansion rate of $1 \times 10^{-6}/°C$. Also, in terms of the crystal strain sensitivity to temperature, PZT materials are much more robust than PMN materials, especially over large temperature (0 °C) variations. However, the strain sensitivity of PMN is less of a factor when used in a laboratory environment, since it approaches the magnitude of the drift of other mechanical components in the system. TRS electrostrictive materials[9] has developed PMN–PT formulations and processing methods that overcome these limitations. PMN promises to deliver ten times the source level produced by current sonar transducers of the same size and weight. Ecertec Electrostrictive Materials[3] of the United Kingdom are also commercially available.

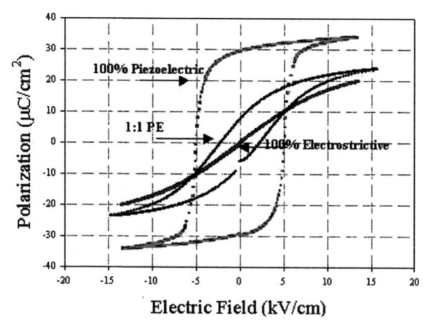

Figure 4.3 Hysteresis in PZT and PMN.
Reproduced from ref. 9 with permission from Springer Nature, Copyright 2005.

Exercise 4.4 What is the difference between the characteristics of piezoelectric materials and electrostrictive materials in terms of hysteresis in polarization?

Answer to Exercise 4.4 According to Figure 4.3, PMN materials have better characteristics than conventional PZT elements. Piezoelectric devices exhibit a hysteresis of 12 to 15% while the hysteresis of the PMN is 2%. Figure 4.3 depicts the difference between the hysteresis in piezoelectrics and electrostrictive materials.

4.6 Conclusions

Electrostrictive materials are very good actuators and sensors and are a nonlinear version of piezoelectric materials. It can be concluded that electrostrictive materials are superior in terms of hysteresis and creep compared to piezoelectric materials. Electrostrictive actuators do provide open-loop positioning repeatability improvements of an order of magnitude or more compared to piezoelectric materials. These actuators offer a minimum incremental motion of as fine as 0.02 μm. The performance improvement of electrostrictive actuators is due to the characteristics of PMN, a ferroelectric material with superior properties for use in motion control applications. Electrostrictive materials are ferroelectric crystals that deform in proportion to an applied voltage. Unlike those of PZT, PMN ceramics are not poled, therefore, alternating between negative or positive voltage results in elongation in the direction of the applied field, regardless of polarity. PMN is an inherently more stable material since it is not poled like piezoelectrics and its creep is reduced to just 3%, while this climbs up to 15% in PZT devices. As far as actuation under an electric field is concerned, electrostrictive materials such as PMN lead to larger displacement under the same electric field compared to PZTs.

Homework Problems

Homework Problem 4.1

An electrostrictive actuator utilizing a 2 mm thick layer of PVDF can be used for applications requiring very short strokes. What is the strain?

Homework Problem 4.2

Compare the constitutive equations for electrostrictive materials with those of piezoelectric materials.

References

1. R. E. Newnham, V. Sundar, R. Yimnirun, J. Su and Q. M. Zhang, Electrostriction: Nonlinear Electromechanical Coupling in Solid Dielectrics, *J. Phys. Chem. B*, 1997, **101**, 10141–10150.
2. Electrostrictive Materials from TRS Technologies, 2018, http://www.trstechnologies.com/Materials/electrostrictive_materials.php.
3. Ecertec Electrostrictive Materials, from Ecertec Ltd., http://www.ecertec.com/electro.htm, 2018.
4. V. Giurgiutiu and S. E. Lyshevski, *Micromechatronics: Modelling, Analysis, and Design with MATLAB*, CRC Press, London, New York, 2004.
5. M. Hu and H. Du, *et al.*, Motion control of an electrostrictive actuator, *Mechatronics*, 2004, **14**(2), 153–161.
6. D. Damjanovic and R. E. Newnham, Electrostrictive and Piezoelectric Materials for Actuator Applications, *J. Intell. Mater. Syst. Struct.*, 1992, **3**(2), 190–208.
7. L. Fabiny and S. T. Vohra, *et al.*, High-resolution fiber-optic voltage sensors based on the electrostrictive effect, *Ferroelectrics*, 1994, **151**(1), 91–96.
8. M. Lallart, P.-J. Cottinet, D. Guyomar and L. Lebrun, Electrostrictive Polymers for Mechanical Energy Harvesting, *J. Polym. Sci., Part B: Polym. Phys.*, 2012, **50**, 523–535.

9. A. Hall, M. Allahverdi, E. K. Akdogan and A. Safari, Development and Electromechanical Properties of Multimaterial Piezoelectric and Electrostrictive PMN-PT Monomorph Actuators, *J. Electroceram.*, 2005, **15**(2), 143–150.

10. B. K. Mukherjee, S. Sherrit, and G. Yang, The Characterisation of Piezoelectric and Electrostrictive Materials for Underwater Acoustic Transducers, Royal Military College of Canada, Report No. DREA-SR-1999-162-PAP-36 — CONTAINED IN CA000150, 1999.

11. B. K. Mukherjee and S. Sherrit, Characterisation of Piezoelectric and Electrostrictive Materials for Acoustic Transducers: I. Resonance Methods, Invited Paper, *Fifth International Congress on Sound and Vibration*, Adelaide, South Australia, December 15–18, 1997.

5 Review of the use of Fibrous Contractile Ionic Polyacrylonitrile (PAN) in Smart Materials and Artificial Muscles

Mohsen Shahinpoor

Mechanical Engineering Dept., University of Maine, USA
Email: shah@maine.edu

5.1 Introduction

Polyacrylonitrile (PAN) fibers in an active form (PAN or PAN gel modified by annealing/cross-linking and partial hydrolysis) elongate and contract when immersed in pH solutions (caustic and acidic solutions, respectively).[1–16] Activated PAN fibers can also contract and expand in polyelectrolyte when electrically and ionically activated with cations and anions, respectively. The change in length for these pH-activated fibers is typically greater than 100%. However, more than 900% contraction/expansion of PAN nanofibers (less than 1 micron in diameter) have been observed in our laboratories. PAN muscles present great potential as artificial muscles for linear actuation, as depicted in Figure 5.1(a) and (b) below.

Fundamentals of Smart Materials
Edited by Mohsen Shahinpoor
© The Royal Society of Chemistry 2020
Published by the Royal Society of Chemistry, www.rsc.org

(a)

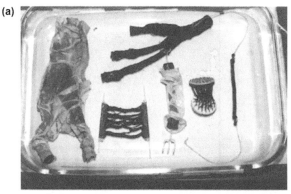

(b)

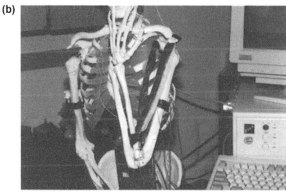

Figure 5.1 Various possible configurations for (a) PAN artificial muscles and (b) their use in an antagonist bicep–tricep configuration in a skeletal arm.

5.2 Ionic PAN Fibers in General

As can be seen in Figure 5.2, the change in length of these pH-activated fibers is typically greater than 100% contraction/expansion of PAN.

The basic unit of commercially available PAN fibers (Mitsubishi Rayon Co., Japan, Orlon) that can be properly handled from an engineering point-of-view is a "single strand". One PAN strand consists of approximately two thousand filaments, where the typical diameter of each filament is approximately 10 µm in a raw state and 30 µm in the fully elongated state (gel).

The advantages of PAN among other electrolyte gels are that PAN gel fibers have good mechanical properties, which compare to those of biological muscles. The large volume change in a PAN fiber gel also allows a reduction in the size of the gel, which is an important factor

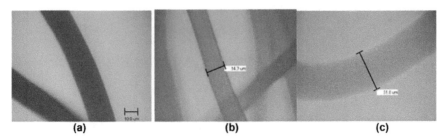

(a)   (b)   (c)

Figure 5.2 PAN fibers in different states (a): oxidized PAN (prior to activation); (b): at low pH contracted PAN (1N HCl); (c); at high pH
expanded PAN (1N LiOH).
Reproduced from ref. 16 with permission from the Royal Society
of Chemistry.

in determining response time. This procedure results in the need to
study PAN fibers and makes the material quite promising for use as a
high-performance artificial muscle. PAN fibers can convert chemical
energy directly into mechanical motion. Figure 5.3[16] depicts a possible explanation for the contraction and expansion of modified PAN
fibers. Based on ion diffusion theory, the response time of swelling is
proportional to the square of the gel fiber diameter. Surface/volume
ratio also affects the response time. Note that such pH-induced
contraction–expansion of modified PAN fibers can also be induced
electrically in a chemical cell by electrolysis and production of H^+ and
OH^- ions.

By annealing the raw PAN fibers (Orlon) at temperatures of
220–240 °C, pyridine rings are formed by crosslinking. The cross-
linked PAN fibers can then be made active[1–16] by boiling the fibers in
1N NaOH or LiOH solutions (saponification of the annealed PAN
fibers).

The annealing temperature and time will determine the degree of
crosslinking. Typical stress–strain curves for active PAN (swollen and
contracted) fiber bundles is also depicted in Figure 5.4.

A possible molecular structure for activated PAN fibers is shown in
Figure 5.5, as discussed by Umemoto, Okui, and Sakai[3] and Hu.[4]
Commonly, one fiber strand has about 2000 microfibers. Each mi-
crofiber has an approximate diameter of 10 μm.

5.3 Preparation of C-PAN

The raw PAN fibers manufactured by Mitsubishi are known as ORLON
or artificial silk. PAN fibers can be activated by first crosslinking them
by sintering in a temperature field. Thus, they are first annealed in air

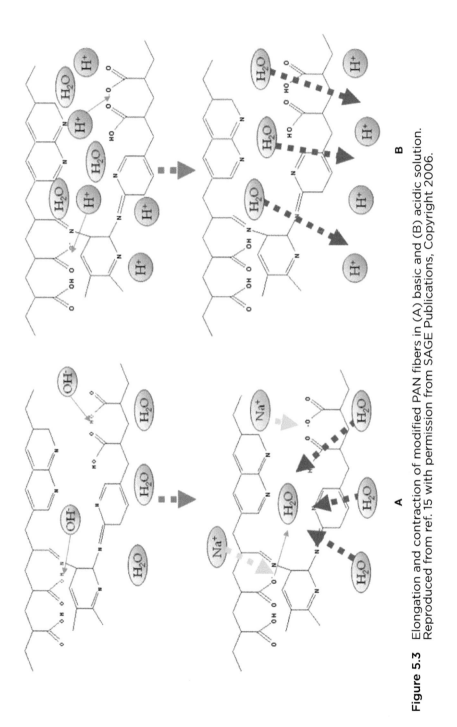

Figure 5.3 Elongation and contraction of modified PAN fibers in (A) basic and (B) acidic solution. Reproduced from ref. 15 with permission from SAGE Publications, Copyright 2006.

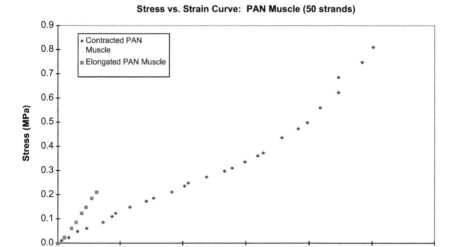

Stress vs. Strain Curve: PAN Muscle (50 strands)

Contracted muscle: length under 0 load = 6.9 cm; maximum load applied = 650 g (did not break); max length = 24.2 cm. Elongated muscle: Length under 0 load = 15.5 cm; maximum load applied = 200 g (broke under load); max length = 20.3 cm.
Estimated cross-sectional area of one strand = 1.57×10^{-7} m^2; for 50 strands = 7.85×10^{-6} m^2.

Figure 5.4   Normal stress–strain relationship for the contracted and the expanded state of PAN fibrous muscles.

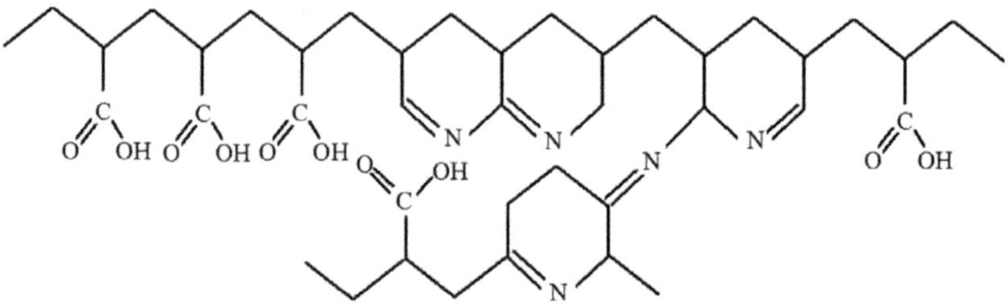

Figure 5.5   Activated crosslinked PAN.

in a temperature range of 220–240 °C for around 2 h. The pre-oxidized PAN fibers are dark brown or black depending upon the level of crosslinking. Second, they are saponified in an alkaline solution (boiling in a 1N NaOH or LiOH solution for 20–30 minutes). After saponification in NaOH or LiOH, the fibers actively contract in a H^{+} or acidic pH environment. After activation, the PAN fibers become highly elastic (hyperelastic), like a rubber band. The elongation and contraction behavior of PAN fibers is interesting in the sense that hysteresis exists upon changing the pH value of activated PAN fibers.

Based upon the Donnan theory of ionic equilibrium,[1] important forces arise from the induced osmotic pressure of free ions between activated PAN fibers and their environment. Here, the ionic interaction of fixed ionic groups and the network itself are at work and induce osmotic pressure of free ionic groups as the dominant force.

5.4 Force–Strain Variations in Active PAN Fibers[16]

Figure 5.6a and b depict the variations in strain with the force of active elongated and contacted PAN fibers, respectively, similar to as in Figure 5.4.[16]

PAN fibers had initial lengths of about 175 mm in an expanded state and shrunk to 110 mm in an acidic environment.

5.5 Variations in the Length of Modified PAN Fibers *versus* the pH Variations of the Solutions in Which they are Contained

The variation in the PAN fiber length *versus* the pH is shown in Figure 5.7.[16] The active PAN fiber length starts to change in the pH range of 10–12 and then a drastic change is observed upon an increase in the pH to pH 12–13. The same drastic change in length is also observed in the pH range of 4–2 when the pH decreases.

Figure 5.8 shows profiles of force development for both the actuation systems.[16] Single strand fibers produced an approximate force of 0.1 N for both activation methods and had similar standard deviation ranges.

5.6 Effect of Different Anions on the Generative Force Characteristics

Ions residing in the molecular structure of the PAN polymer and the solvent are part of the reason behind the generation of the osmotic pressure that swells and shrinks the PAN gel. It was generally noticed that the difference in force generation of the three acidic solutions was very small or negligible.

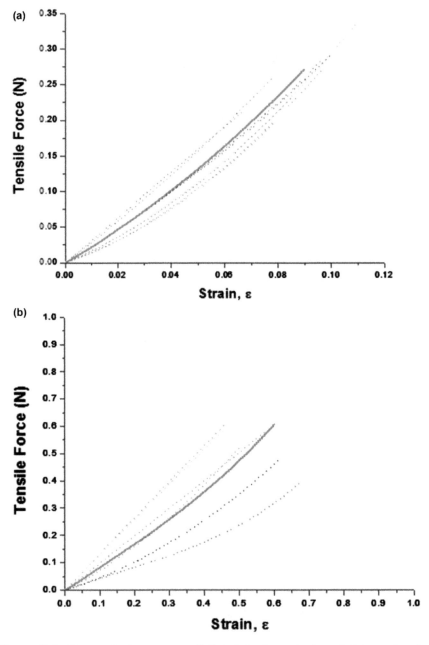

Figure 5.6 Force–strain curves of the (a) elongated and (b) contracted state of a single PAN strand. The solid line is the average value ten samples. The dotted lines show the force–strain curves of each sample. Large scattering between different samples is observed, which is probably due to the non-uniformity of the fibers. Reproduced from ref. 17 with permission from Elsevier, Copyright 2006.

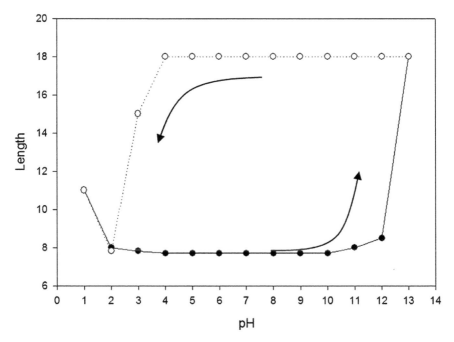

Figure 5.7 Length changes of the PAN fibers *versus* a variation in pH. Large hysteresis was observed upon a pH change from low to high values and from high to low values.
Reproduced from ref. 17 with permission from Elsevier, Copyright 2006.

5.7 Generative Force Characteristics: Effect of Acidity

The PAN fibers were saponified by boiling them in a 1 M LiOH or NaOH solution for 30 minutes. HCl solutions with different concentrations (0.01, 0.1, 0.5, 1, and 2 M) were also prepared. The PAN fiber contractile force generated in each HCl acidic solution was then measured. In Figure 5.9, the measured generative force (contractile force) for different acid concentrations is presented.[16] Force generation generally increases as the concentration of the acidic solution increases.

5.8 Performance of a PAN Bundle Artificial Muscle

Chemically activated PAN artificial muscle systems require a circulatory system of pH solutions like the blood circulatory system in mammalian bodies. This system supplies acidic and basic solutions to the PAN muscle and allows drainage of the waste liquid out of the

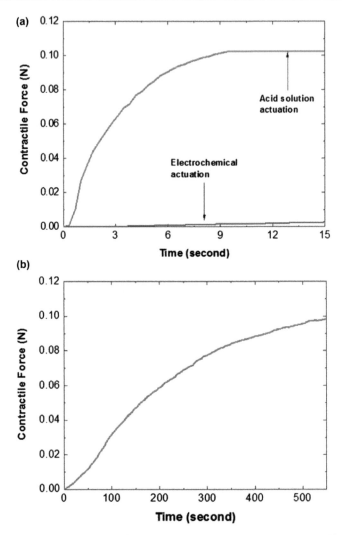

Figure 5.8 Profile of single strand force generation in (a) 1 M HCl and (b) a
5 V electric field. Note that the slow electrochemically induced
swelling correlates with the time dependence of the electrochemically induced pH change.
Reproduced from ref. 17 with permission from Elsevier, Copyright
2006.

actuation system. An electrochemically driven actuation system may
be a more convenient system to provide circulation.

This section describes the measured performance of chemically
and electrochemically induced PAN bundle artificial muscles. The use
of bundles rather than a strand (~2000 fibers) increases the practi-
cality of employing PAN artificial muscles.

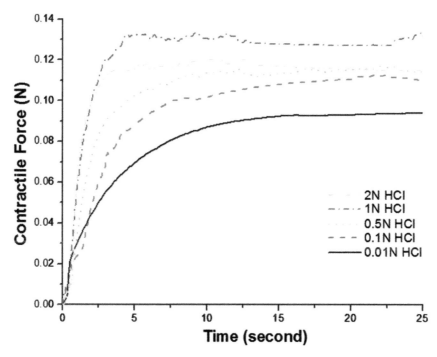

Figure 5.9 Variation in the force generated by active PAN fibers in various pH concentrations of HCl acid solutions with fibers of 100 mm in length.
Reproduced from ref. 16 with permission from the Royal Society of Chemistry.

5.9 Electrical Activation of Conductive PAN (C-PAN) Muscles

Electrical activation of active PAN fibers is also possible and is performed in an electrochemical cell with a saline solution, as shown in Figure 5.10.

PAN fibers can be activated electrically, by putting a conductive medium in contact with or within the PAN fibers, in which a conductive phase is intermingled. Essentially H^+ ion production by voltage-induced hydrolysis changes the pH of the solution and causes C-PAN fiber contraction or expansion. Graphite fibers can be intermingled with PAN fibers and the combined fiber bundle used as the anode or cathode in an electrochemical cell. Upon being hydrogenated in the vicinity of the C-PAN anode, the decrease in the pH causes the C-PAN fibers to contract *via* the same effect as chemical activation upon a variation in the pH. Also, reversing the polarity of

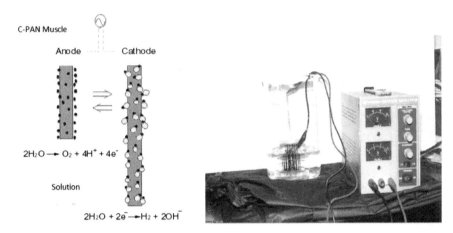

Figure 5.10   Electrochemical activation of C-PAN artificial sarcomeres and muscles. (a) The PAN fibers contracting at the anode due to the production of H^+ and expanding at the cathode due to the production of OH^-. This phenomenon describes the operating principles of the conducting C-PAN fibers intermingled with electrically conductive fibers such as graphite.

the voltage in the electrochemical cell results in the elongation of PAN fibers in the electrochemical cell.

5.10   Electric Current Effect on Force Generation

If the current densities are around 10 and 20 mA cm^{-2}, the C-PAN bundle actuators are under 10 gm$_f$ pretension stress. Figure 5.11 shows the force generation under varying current densities and voltage changes.[16] There is usually a 250–300 second time interval before the contractile force is developed, and as the current density increases, the time interval decreases.

5.11   Mathematical Modeling of the Contraction and Swelling of Active PAN Muscles

Hydrogen bonding between neighboring carboxylic acid groups in PAN muscles[9–12] leads to the contraction and exudation of water from the PAN muscles. On the other hand, at a higher pH, the PAN muscles expand by taking water inside the molecular network.

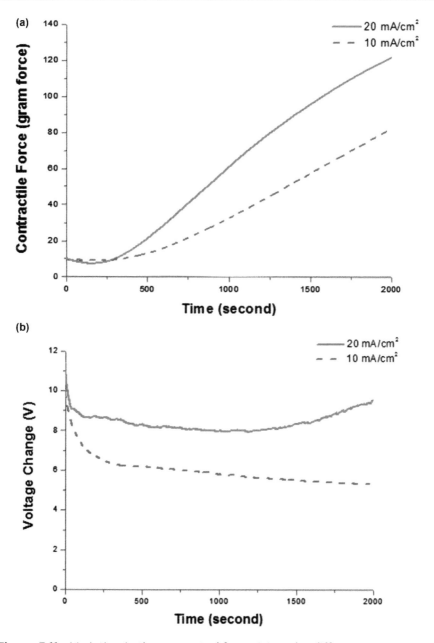

Figure 5.11   Variation in the generated force (a) under different currents and (b) different voltages 10 gm$_f$ pre-tension.
Reproduced from ref. 15 with permission from SAGE Publications, Copyright 2006.

This degree of hydration is related to the degree of swelling or V/V_0, where V is the current volume of PAN muscles, and V_0 is the initial volume of the PAN muscles. The Donnan theory of ionic equilibrium,[6,16] establishes that the induced osmotic pressure of free ions between activated PAN fibers and their environment, the ionic interaction of fixed ionic groups and the network structure are the important driving forces.

If we describe the kinetics of PAN fibers using diffusion controlled slab-type gels, then the elongation is:[7]

$$\frac{\Delta l}{\Delta l_0} = 1 - \sum_{n=0}^{\infty} \frac{8}{(2n+1)^2 \pi^2} \exp\left(-\frac{(2n+1)^2 t}{\tau}\right) \tag{5.1}$$

where the characteristic time, τ, is given by

$$\tau = \frac{4 l_{ch}^2}{\pi^2 D} \tag{5.2}$$

5.12 Modeling of the Expansion and Contraction of PAN Muscles Based on Electrocapillary Effects

Considering the gel fiber to be a swollen cylinder with an outer radius, r_0, and an inner radius, r_i, and assuming the electric field to be aligned with the long axis of this cylindrical macromolecule ionic chain, then the following equation, the standard conservation of linear momentum of the dynamic PAN muscles in the presence of liquid solvent absorption (expansion, dilution, swelling) or extraction (contraction, de-swelling) can be derived:

$$\rho \frac{dv}{dt} = \rho g + \rho^* E + \mu \nabla^2 v - \nabla p \tag{5.3}$$

Here, ρ is the density of the liquid solvent (assumed to be incompressible), v is the 3D liquid velocity vector, ∇ is the gradient vector operator, ∇^2 is the Laplacian operator, g is the local gravitational acceleration vector, μ is the solvent viscosity, p is the hydrostatic pressure, E is the imposed electric field vector and ρ^* is the charge density. The charge density ρ^* is governed by Poisson's equation:

$$\rho^* = -D \nabla^2 \psi \tag{5.4}$$

Here, in eqn (5.4), D is the dielectric constant of the liquid phase and ψ is governed by the following Poisson–Boltzmann equation:

$$\nabla^2\psi = (4\pi n\varepsilon/D)\,\exp\left[-\varepsilon\psi/kT\right] \tag{5.5}$$

where n is the number density of counter ions, ε is their average charge, k is the Boltzmann constant and T is the absolute temperature. The electrostatic potential in polyelectrolyte solutions for fully stretched macromolecules is given by the following equation, which is an exact solution to the Poisson–Boltzmann eqn (5.2–5.4) in cylindrical coordinates:

$$\psi(r,t) = [kT/\varepsilon]\ln\{[r^2/(r_0^2 - r_i^2)]\,\sinh^2[\beta\,\ln(r/r_0) - \tan^{-1}\beta]\} \tag{5.6}$$

where β is related to $\lambda = (\alpha\varepsilon^2/4\pi DbkT)$, α is the degree of ionization (*i.e.*, $a = n/Z$, where n is the number of polyions and Z is the number of ionizable groups), and b is the distance between the polyions in the network. Furthermore, $n = [\alpha\varepsilon^2/(4\pi DbkT)]$ and the β values can be determined using the following equation:

$$\lambda = \frac{1 - \beta^2}{1 + \beta\coth[\beta\ln(r_0/r_i)]} \tag{5.7}$$

Let us further assume that due to cylindrical symmetry, the velocity vector $v = (v_r, v_\theta, v_z)$ is such that only v_z depends on r and further that $v_\theta = 0$. Thus, the governing equations for $v_z = v$ reduce to:

$$\rho\left(\frac{\partial v}{\partial t}\right) = f(r,t) + \mu\left(\frac{\partial^2 v}{\partial r^2} + \frac{1}{r}\frac{\partial v}{\partial r}\right) - \frac{\partial p}{\partial r} \tag{5.8}$$

Let us assume a negligible radial pressure gradient and assume the following boundary and initial conditions:

At $t = 0$, $r_i \le r \le r_o$, $v = 0$, at $r = r_i$, $\forall t$, $v(r_i) = 0$, and at $r = r_o$, $\forall t$, $(\partial v/\partial r)_{r=r_o} = 0$. Furthermore, the function $f(r,t)$ is given by:

$$f(r,t) = n\varepsilon E(r,t)\{[k^2 r^2/2\beta^2]\,\sinh^2\,[\beta\ell n(r/r_o) - \tan^{-1}\beta]\}^{-1} \tag{5.9}$$

where $k^2 = (n\varepsilon^2/DkT)$.

An exact solution to the given set of equations can be shown to be:

$$v(r,t) = \sum_{m=1}^{\infty} e^{-(\mu/\rho)\beta^2 mt} k_0(\beta_m r) \int_0^t e^{(\mu/\rho)\beta_m^2 \xi} A(\beta_m, \xi)\,d\xi \tag{5.10}$$

where β_m values are the positive roots of the following transcendental equations:

$$\frac{J_0(\beta\,r_i)}{J_0'(\beta\,r_0)} - \frac{Y_0(\beta\,r_i)}{Y_0'(\beta\,r_0)} = 0 \tag{5.11}$$

where J_0, Y_0, J_0', Y_0', are the Bessel functions of zero order of the first and second kind and their derivatives evaluated at r_0, respectively, and

$$k_0(\beta_m, r) = N^{-(1/2)}\left\{\frac{J_0(\beta r)}{\beta_m J_0'(\beta r_0)} - \frac{Y_0(\beta r)}{\beta_m Y_0'(\beta r_0)}\right\} = N^{-(1/2)}R_0(\beta_m, r) \tag{5.12}$$

where $N = (r_0/2)R_0^2(\beta_m, r_0) - (r_i^2/2)R_0'^2(\beta_m, r_i)$, $A(\beta_m, \xi) = (1/\mu\rho)\int_{r_i}^{r_0} \varsigma\,k_0(\beta_m, \varsigma)f(\varsigma, \xi)d\varsigma$

Considering the ratio $W(t)/W(0)$, where $W(t)$ is the weight of the entire gel at time t, and $W_0 = W(0)$ is the weight of the gel at time $t = 0$, just before the electrical activation. Thus,

$$W(t) = W_0 - \int_0^t\int_{r_i}^{r_0} 2\pi\rho v(r, t)rdrdt \tag{5.13}$$

can be simplified to

$$[W(t)/W_0] = 1 - W_0^{-1}\int_0^t\int_{r_i}^{r_0} 2\pi\rho v(r, t)rdrdt \tag{5.14}$$

The initial weight of the gel is related to the initial degree of swelling $q = V(0)/V_p$, where $V(0)$ is the volume of the gel sample at

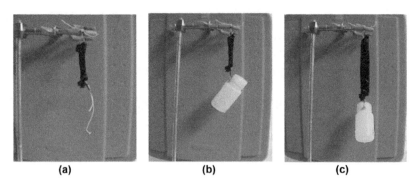

(a) (b) (c)

Figure 5.12 Degree of swelling of bundles of 20 strands of annealed ORLON fibers, (a) pH$=7$ (neutral), (b) pH$=1$, acidic (contracted), (c) pH$=14$, basic (expanded).

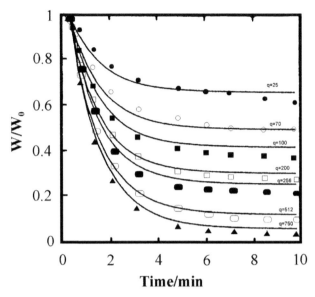

Figure 5.13 Computer simulation (solid lines) and experimental results (scattered points) for the time profiles of the relative weights of the gel samples for various degrees of swelling, q.

$t = 0$ and V_p is the volume of the dry polymer sample. Figure 5.12 depicts the PAN muscles at (a) pH $= 7$, (b) pH $= 1$ and (c) pH $= 14$ as the acidity of the PAN fiber bundle is changed by increasing the degree of swelling.

Assuming the cross-section of the gel remains unchanged during contraction and if $\varepsilon = e = 1.6 \times 10^{-19}$ Coulombs, $T = 300$ K, $\alpha = 1$, $D = 80$, $\mu = 0.8 \times 10^{-3}$ Pa s, $\rho = 1000$ Kg m^{-3}, $b = 2.55 \times 10^{-10}$ m, $k = 1.3807 \times 10^{-23}$ Joules K^{-1}, $r_i = 6.08 \times 10^{-10}$ m, $r_o = r_i q^{(1/2)}$, and $q = 25, 70, 100, 200, 256, 512, 750$. Computer calculations were performed and the results are shown in Figure 5.13.

This section completes our numerical results and comparison with experimental results for the contraction of PAN fibers.

5.13 Conclusions

Active PAN fibers were shown to be capable of contraction and expansion in different pH solutions and generating contractile forces comparable to mammalian muscles. The possibility of electro-chemically activating them was also discussed.

Mathematical formulation and exact solutions to the degree of swelling, as related to the degree of contraction and expansion of PAN muscles were also presented. Numerical and experimental results about the degree of PAN fiber contractions and the associated contractile force generations were also discussed and numerically presented. The chapter also discussed PAN conductor composites (C-PAN) as presenting great potential for fabricating electrically actuated contractile PAN muscles.

Homework Problems

Homework Problem 5.1

Propose a design for a resilient PAN actuator.

Homework Problem 5.2

Describe the pH-activated contraction and expansion mechanisms of PAN.

Homework Problem 5.3

Describe the differences between the stress–strain curves of contracted and expanded PAN fibrous muscles.

References

1. P. Flory, *Principles of Polymer Chemistry*, Cornell University Press, Ithaca, New York, 1953.
2. R. P. Hamlen, C. E. Kent and S. N. Shafer, Electrolytically Activated Contractile Polymer, *Nature*, 1965, **206**, 1149–1150.
3. S. Umemoto, N. Okui and T. Sakai, Contraction Behavior of Poly(acrylonitrile) Gel Fibers, *Polymer Gels*, ed. D. DeRossi, *et al.*, Plenum Press, New York, 1991, pp. 257–270.
4. X. Hu, Molecular Structure of Polyacrylonitrile Fibers, *J. Appl. Polym. Sci.*, 1996, **62**, 1925–1932.
5. K. Salehpoor, M. Shahinpoor and M. Mojarrad, Electrically Controllable Artificial PAN Muscles, *Proc. SPIE 1996 North American Conference on Smart Structures and Materials*, February 27–29, 1996, San Diego, California, vol. 2716, paper no. 07, 1996.
6. H. B. Schreyer, M. Shahinpoor and K. J. Kim, Electrical Activation of PAN-Pt Artificial Muscles, *Proceedings of SPIE/Smart Structures and Materials/Electroactive Polymer Actuators and Devices*, Newport Beach, California, vol. 3669, pp. 192–198, March 1999.
7. H. B. Schreyer, N. Gebhart, K. J. Kim and M. Shahinpoor, Electric Activation of Artificial Muscles Containing Polyacrylonitrile Gel Fibers, *Biomacromolecules*, 2000, **1**, 642–647.

8. M. Shahinpoor, K. J. Kim and H. B. Schreyer, Artificial Sarcomere and Muscle Made with Conductive Polyacrylonitrile (C-PAN) Fiber Bundles, *Proceedings of SPIE 7th International Symposium on Smart Structures and Materials*, Newport Beach, California, vol. 3687, pp. 243–251, March 2000.
9. K. J. Kim and M. Shahinpoor, Electrical Activation of Contractile Polyacrylonitrile (PAN)-Conductor Composite Fiber Bundles As Artificial Muscles, *Proceedings of the First World Congress On Biomimetics and Artificial Muscle (Biomimetics 2002)*, December 9–11, 2002, Albuquerque Convention Center, Albuquerque, New Mexico, USA, 2002.
10. M. Shahinpoor and M. Ahghar, Modeling of Electrochemical Deformation in Polyacrylonitrile (PAN) Artificial Muscles, *Proceedings of the First World Congress On Biomimetics and Artificial Muscle (Biomimetics 2002)*, December 9–11, 2002, Albuquerque Convention Center, Albuquerque, USA, 2002.
11. K. J. Kim, J. Caligiuri, K. Choi, M. Shahinpoor, I. D. Norris and B. R. Mattes Polyacrylonitrile Nanofibers as Artificial Nano-Muscles, *Proceedings of the First World Congress On Biomimetics and Artificial Muscle (Biomimetics 2002)*, December 9–11, 2002, Albuquerque Convention Center, Albuquerque, New Mexico, USA, 2002.
12. M. Shahinpoor, K. J. Kim, L. O. Sillerud, I. D. Norris and B. R. Mattes, Electro-active Polyacrylonitrile Nanofibers as Artificial Nanomuscles, *Proceedings of SPIE 9th Annual International Symposium on Smart Structures and Materials*, San Diego, California, SPIE Publication No. 4695-42, March 2002.
13. K. J. Kim, J. Caligiuri and M. Shahinpoor, Contraction/Elongation Behavior of Cation-Modified Polyacrylonitrile Fibers, *Proceeding of SPIE 10th Annual International Symposium on Smart Structures and Materials*, 2–6 March 2003, San Diego, California, SPIE Publication No. 5051-23, pp. 207–213, 2003.
14. K. J. Kim, K. Choe, R. Samathan, J. Nam, M. Shahinpoor and J. Adams, Toward Nanobiomimetic Muscles: Polyacrylonitrile Nanofibers, *Proceeding of SPIE 11th Annual International Symposium on Smart Structures and Materials*, 14–18 March, 2004, San Diego, California, SPIE Publication No. 5385-62, pp. 33–43, 2004.
15. K. Choi, K. J. Kim, D. Kim, C. Manford, S. Heo and M. Shahinpoor, Performance Characteristics of Electro-Chemically Driven Polyacrylonitrile Fiber Bundle Actuators, *J. Intell. Mater. Syst. Struct.*, 2006, **17**(7), 563–576.
16. K. J. Kim and K. Choe, Electrochemically Controllable Polyacrylonitrile-Derived Artificial Muscle as an Intelligent Material, in *Intelligent Materials*, Royal Society of Chemistry, Cambridge, 2008, pp. 191–203.
17. K. Choe and K. J. Kim, *Sens. Actuators, A*, 2006, **126**, 165–172.

6 Review of Magnetostrictive (MSMs) and Giant Magnetostrictive Materials (GMSs)

Mohsen Shahinpoor

Mechanical Engineering Dept., University of Maine, USA
Email: shah@maine.edu

6.1 Introduction

The magnetostrictive effect was first discovered in 1842 by James Joule,[1] when observing a sample of nickel change dimensions in a magnetic field. The giant magnetostrictive effect was discovered by A.C. Clark[2] in 1980. Magnetostriction was thus defined as a property of ferromagnetic materials that causes them to change their shape when subjected to a magnetic field. This effect is also known as the **Joule Effect**. The essential mechanism is the change in dimensions resulting from the interactive coupling between an applied magnetic field and the magnetization and magnetic moments of the material's domains or magnetic dipoles.[2,3] Figure 6.1 depicts the basic mechanisms of internal magnetic domains aligning in a magnetic field in magnetostrictive materials under no stress. Figure 6.2 depicts the same effect (Figure 6.1), but with the material initially under some stress. Magnetostriction causes mechanical deformation (length or volume changes) in nearly all ferromagnetic materials when they are

Fundamentals of Smart Materials
Edited by Mohsen Shahinpoor
© The Royal Society of Chemistry 2020
Published by the Royal Society of Chemistry, www.rsc.org

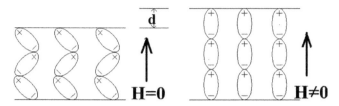

Figure 6.1 Domain movement under an applied magnetic field (Joule Effect).

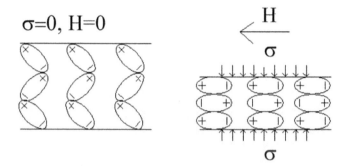

Figure 6.2 Domain movement under applied stress or strain (Villari Effect).

placed in a magnetic field. In the applied magnetic field (H) the domains move or rotate, so that the magnetic poles align, causing a dimensional change (d). This alignment is illustrated in Figure 6.1.

As clearly indicated in Figures 6.1 and 6.2, magnetostriction occurs and leads to dimensional changes in a magnetostrictive sample because of the rotation of small magnetic domains or dipoles within the magnetostrictive material. The resulting magnetic dipole rotation and re-orientation generate internal strains in the material leading to dimensional changes. The relationship between such rotations and resulting strains will be described in the following sections. The magnetostrictive strains in the structure cause dimensional changes, as shown in Figure 6.3. During these dimensional changes, the cross-section of the sample is reduced in a way such that the volume is kept nearly constant. In other words, the volume change is small enough to be negligible under normal operating conditions.

Furthermore, stronger and more definite re-orientation of more domains in the direction of the magnetic field occurs by applying a stronger magnetic field. Saturation sets in when all the magnetic domains have aligned in the direction of the applied magnetic field, as also shown in Figure 6.3a. Typical strain or $\varepsilon = \Delta L/L$ *versus* the applied magnetic field H variations are shown in Figure 6.3a.

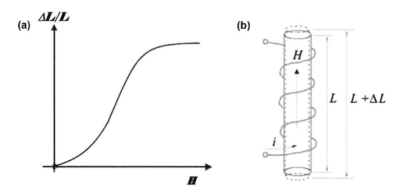

Figure 6.3 Typical linear strain $\Delta L/L$ *versus* a magnetic field H (a) in a magnetostrictive material (b) subject to a current i and magnetic field H.

Figure 6.3b shows the idealized behavior of length change *versus* the applied magnetic field (**Joule effect**). Note that after a while, saturation sets in as there is no additional strain with increasing magnetic field, H, as shown in Figure 6.3a.

It is natural to define the magnetostrictive coefficient, d, as the fractional change in length as the magnetization reaches its saturation value. Ordinary magnetostrictive materials such as cobalt exhibit the largest room temperature magnetostriction of 60 microstrains (strain$\times 10^6$). However, the highest known magnetostriction (0.002 mm^{-1}) is exhibited by Terfenol-D, (Ter for terbium, Fe for iron, NOL for Naval Ordinance Laboratory, and D for dysprosium) discovered in the US Naval Ordnance Laboratory in 1970. The general chemical formula for Terfenol-D is $Tb_xDy_{1-x} Fe_2$, where x is around 0.3.

Magnetostriction occurs because of two distinct processes. In the first process, the domain walls migrate basically due to the externally applied magnetic field, H. Following domain wall migrations, the magnetic dipoles rotate, thus inducing a deformation field within the magnetostrictive material. The deformation is isochoric (constant volume) and is thus accompanied by a deformation in the orthogonal direction. The principal advantage of magnetostrictive materials is that they can produce strains in the order of a few thousand μ strains at moduli, typically around 100 GPa. Commonly, for electrostrictive materials, we observe narrow hysteresis loops. Their low hysteresis and excellent repeatability are significant advantages of magnetostrictive materials.[1-3]

Historically, giant magnetostrictive materials (GMSs) such as rare earth–iron compounds were discovered by A. E. Clark.[2] These rare-earth materials generated magnetostrictive strains that are orders of magnitude larger than the strain observed in nickel.

Exercise Number 6.1: Describe the **Joule effect**.

Answer to Exercise Number 6.1: At the fundamental level, the Joule effect is the change in the dimensions of a ferromagnetic material due to the interactive coupling between an applied magnetic field and the magnetization and magnetic moments of the internal domains or magnetic dipoles. The Joule effect was originally observed in 1842 by James Joule,[1] when observing a sample of nickel.

Exercise Number 6.2: Describe the reverse of the **Joule effect**.

Answer to Exercise Number 6.1: The reverse of the Joule effect is called the **Villari effect**. Here, when the magnetostrictive material is subjected to mechanical stress, its magnetic susceptibility changes due to loading.

6.2 Various Magnetostrictive Effects

Terfenol-D exhibits about 2000 microstrains (0.002 m m^{-1}) in a field of 2 kOe (160 kA m^{-1}) at room temperature and is the most commonly used magnetostrictive engineering material. Magnetostrictive materials shrink in the presence of a magnetic field. Magnetostrictive materials are ferromagnetic with extremely narrow hysteresis loops. They comprise alloys of iron, nickel, and cobalt with rare earth doping. The most effective magnetostrictive material is Terfenol-D, an alloy of terbium, dysprosium, and iron. The name combines terbium, iron, and the Naval Ordinance Laboratory (currently the Naval Surface Weapons Center) where the alloy was invented.[4-6] Terbium is arguably the rarest of the rare earth metals and so Terfenol-D is also the most expensive of the magnetostrictive materials. There are other related effects that should be mentioned here. One effect is called the **Villari Effect**, which is the reverse of the Joule Effect. Here, when the magnetostrictive material is subjected to mechanical stress, its magnetic susceptibility changes. Magnetic susceptibility is usually indicated by the chi symbol, χ, and is equal to the ratio of the magnitude of the internal magnetic polarization, m_p, to the imposed magnetic field strength, B. Note that the magnetic susceptibility, χ, can be both positive and negative. Normal ferromagnetic materials have positive susceptibility, while diamagnetic materials such as water or biological tissues have negative susceptibility. There are two other magnetostrictive related effects,[3] which are called the **Matteucci effect** and the **Wiedemann effect**, respectively. The Matteucci effect manifests itself as helical anisotropy of the magnetic susceptibility when a magnetostrictive material is subjected to a twisting torque or torsional stress. On the other hand, the Wiedemann effect is the torsional deformation of the magnetostrictive material under a helical magnetic field.

Olabi and Grunwald[3] discussed the design and applications of magnetostrictive materials in terms of the various effects emanating from the **Joule effect**, such as the **Villari effect**, the **Wiedemann effect** and the **Matteucci effect**.[1–4] There is another effect called the ΔE-**effect**,[1–4] which changes the Young's modulus of elasticity, E, of the GMS as a result of an applied magnetic field. The $\Delta E/E$ of Terfenol-D is in the range of more than 5 and can be employed in tunable vibration and broadband sonar systems.[1–4] Due to the change in the Young's modulus, there is a change in the velocity of sound inside magnetostrictive materials, and this can be observed. The **inverse Wiedemann effect** is called the **Matteucci effect**.[3] Alternating current fed to a coil creates a longitudinal magnetic field in a sample, and this, in turn, creates a magnetic flux density in the sample. An additional magnetostrictive effect is the **Barret effect**,[3,4] where the volume of GMSs can change if placed in a magnetic field. There is also the inverse Barret effect known as the **Nagaoka–Honda effect**.[3,4] In this effect, if the GMS is placed in a hydrostatic pressure environment, which may cause its volume to decrease, then the magnetic state of the GMS changes accordingly. The most widely employed magnetostrictive effects are the **Joule effect** and its inverse, the **Villari effect**.

6.3   Terfenol-D Availability

As far as giant magnetostrictive Terfenol-D is concerned, Etrema Products Inc.,[7] which is a subsidiary of *Edge Technologies, Inc.,* Ames, IA 50010, www.etrema-usa.com, manufactures solid giant magnetostrictive Terfenol-D powder composites (GMPCs) and thin films.[7] Box 6.1 displays the typical properties of Terfenol-D, as tabulated by Etrema Products Inc.

6.4   Properties of Terfenol-D

Since becoming commercially available in 1987, Terfenol-D is a very popular rare earth material as it presents the best compromise between a large magneto strain and a low magnetic field at

---

Box 6.1   Some typical properties of Terfenol-D[7]

Nominal Composition: Tb $_{0.3}$ Dy $_{0.7}$ Fe $_{1.92}$ GMPC
Mechanical Properties: Young's Modulus $= 25\text{--}35$ GPA, Sound Speed $= 640\text{--}1940$ m s^{-1}, Tensile Strength $= 28$ Mp, Compressive Strength $= 700$ Mpa
Thermal Properties: Coefficient of thermal expansion $= 12$ ppm/°C, Specific Heat $= 0.35$ kJ/kg-K, Thermal Conductivity $= 13.5$ W/m-k
Electrical Properties: Resistivity $= 58 \times 10^{-8}$ Ohm-m, **Curie Temperature** $= 380$ °C
Magnetostrictive Properties: Strain ($\sim$linear) $= 800\text{--}1200$ ppm, Energy Density $= 14\text{--}25$ kJ m^{-3}
Magnetomechanical Properties: Relative Permeability $= 3\text{--}10$, Coupling Factor $= 0.75$

---

room temperature. Positive magneto strains of up to 2000 ppm (parts per million) (0.002 m m^{-1}) under magnetic fields of 100 to 200 kA m^{-1} have been reported.[7] The following relationships in terms of the magnetization, M, magnetic susceptibility, χ_{m}, relative permeability, μ_{r}, and magnetic flux density, B, are commonly used in designing high power actuators and transducers:

$$\chi_{m} = M/H \tag{6.1}$$

$$B = \mu_{0} (H + M) \tag{6.2}$$

where μ_{0} is defined as the magnetic permittivity of free space and H is the applied peripheral magnetic field strength.

If ferromagnetic materials are heated above what is called the Curie temperature, they become paramagnetic. The Curie temperature or T_{c} is a temperature beyond which ferromagnetic materials lose their magnetic properties. Beyond the Curie temperature, the magnetic ordering of the dipoles of the ferromagnetic material is deteriorated by the thermal energy $E_{thermal}$. Note that $E_{thermal} = k_{B} T$ and Curie's Law is given by $\chi_{m} = C/T$, where k_{B} is the Boltzmann constant $= 1.38064852(79) \times 10^{-23}$ J K^{-1}, T is the temperature in Kelvin, χ_{m} is the magnetic susceptibility, and C is the Curie Constant.

The Curie temperatures of some ferromagnetic materials are 1043 K for iron, 627 K for nickel, 1338 K for cobalt, and 293 K for gadolinium. Note that ferromagnetism is the most dominant magnetic property compared with other forms of magnetism. Ferromagnetic materials are heavily used in many applications in engineering, industry, science, education, magnetic storage devices, electromagnets, transformers, electric motors, generators, control devices and actuation, and sensing systems. Actuation, energy harvesting, and sensing are just additional emerging applications, especially in connection with magnetostrictive materials.

Exercise 6.2 Describe the Villari effect.

Answer to Exercise 6.2: The Villari effect is the inverse Joule effect. Upon applying stress on a magnetostrictive material, the magnetic susceptibility is changed.

Other magnetostrictive phenomena[1–3] are the torsional motion in a material (Wiedemann effect) caused by an external magnetic field and the inverse Wiedemann effect, the so-called **Matteucci effect**, where a torsional deformation of the sample creates a voltage along the length of the material.

6.5 GMS Constitutive Equations

Constitutive modeling of GMSs such as Terfenol-D should describe the characteristics of actuation and sensing in GMSs based on the interaction between the mechanical and magnetic energy of magnetostrictive materials. Here, we follow the approach of Olabi and Grunwald.[3] Eqn (6.3) and (6.4) are the linear magnetostrictive constitutive equations proposed in ref. 3:

$$\mathbf{S} = (1/E^H)\,\mathbf{T} + \mathbf{d}\,\mathbf{H} \tag{6.3}$$

$$\mathbf{B} = \mathbf{d}\,\mathbf{T} + \mathbf{\mu}^T\,\mathbf{H} \tag{6.4}$$

where **S** is the strain tensor, **T** is the stress tensor, **E** is the modulus of the elasticity tensor at a constant applied magnetic field strength **H**, **d** is the matrix of magnetostrictive coefficients, $\mathbf{\mu}^T$ is the magnetic permeability at a constant stress, **T**, **H** is the applied magnetic field strength vector and **B** is the magnetic flux density vector within the Terfenol-D material. The above eqn (6.3) and (6.4) are the simplest expressions for the strain, **S**, and magnetic flux density **B**.

Note how they can be linearly dependent on the stress, **T**, and the applied magnetic field, **H**. These equations clearly show the interplay between the stress tensor, **T**, the strain tensor, **S**, the applied magnetic field, **H**, the magnetostrictive coefficients matrix **d**, and magnetic induction flux density tensor, **B**, in moving and aligning magnetic domain walls within the material. Eqn (6.3) and (6.4) describe the response of the GMS material to externally applied conditions. Increasing the applied field, **H**, or magnetic permeability, $\mathbf{\mu}^T$, increases the magnetic induction, **B**, if the stress is kept constant. On the other hand, **B** decreases if the stress, **T**, decreases while **H** is kept constant.

Exercise Number 6.3: Calculate the induced magnetic field strength in a coil of n turns carrying a current I.

Answer to exercise number 6.3: The magnetic field strength, H, can be calculated using the relationship $B = \mu H = \mu n I$ or $H = nI$, with n being the number of turns, I is the current in Amps, and μ_0 is the magnetic permeability of the free space, $\mu_0 = 4\pi \times 10^{-7}\ NA^{-2}$.

Exercise Number 6.4: Discuss the constitutive equations for giant magnetostrictive materials.

Answer to exercise number 6.4: The linear magnetostrictive constitutive equations are given by eqn (6.3) and (6.4):

$$\mathbf{S} = (1/\mathbf{E}^{\mathbf{H}})\ \mathbf{T} + \mathbf{d}\ \mathbf{H} \qquad (6.3)$$

$$\mathbf{B} = \mathbf{d}\ \mathbf{T} + \mathbf{\mu}^{\mathbf{T}}\ \mathbf{H} \qquad (6.4)$$

where $\mathbf{S}$ is the strain tensor, $\mathbf{T}$ is the stress tensor, $\mathbf{E}$ is the modulus of elasticity tensor at a constant applied magnetic field strength $\mathbf{H}$, $\mathbf{d}$ is the matrix of magnetostrictive coefficients, $\mathbf{\mu}^{\mathbf{T}}$ is the magnetic permeability at a constant stress $\mathbf{T}$, $\mathbf{H}$ is the applied magnetic field strength vector and $\mathbf{B}$ is the magnetic flux density vector within the Terfenol-D material.

6.6 Conclusions

This chapter presented a brief description of MSMs and GMSs. The advantages of magnetostrictive applications are that they are an excellent means for high volume applications for the implementation of many applications and technologies. Because some active GMSs such as Terfenol-D have been developed, giant magnetostrictors with fairly stable characteristics over a wide range of temperatures are being designed and developed. Magnetostrictive technology has been successfully employed in defense as well as several industries, such as the automotive industry. For high volume applications and the production of GMSs, the appropriate manufacturing technologies have also been developed. GMSs are now are also being used in different disciplines such as the medical industry, in biomedical implants, and in the automotive industry.

Homework Problems

Homework Problem 6.1

Describe the Matteucci effect.

Homework Problem 6.2

Describe the Wiedemann effect.

Homework Problem 6.3

Describe the ΔE-effect.

Homework Problem 6.4

Describe the Barret effect.

Homework Problem 6.5

Describe the Nagaoka–Honda effect.

References

1. J. Joule, *Sturgeon's Ann. Electr.*, 1842, **8**, 219.
2. A. E. Clark, Magnetostrictive rare earth-Fe$_2$ compounds, in *Chapter 7, Handbook of Ferromagnetic Materials*, vol. 1, North-Holland, Amsterdam, 1980, pp. 531–589.
3. A. G. Olabi and A. Grunwald, Design and application of magnetostrictive materials, *Mater. Des.*, 2008, **29**(2), 469–483.
4. M. B. Moffett, *et al.*, Characterization of Terfenol-D for magnetostrictive transducers, *J. Acoust. Soc. Am.*, 1991, **89**(3), 1448–1455.
5. L. Sandlund, *et al.*, Magnetostriction, elastic moduli, and coupling factors of composite Terfenol-D, *J. Appl. Phys.*, 1994, **75**(10), 5656–5658.
6. T. Ueno, J. Qiu and J. Tani, Magnetic force control with a composite of giant magnetostrictive and piezoelectric materials, *IEEE Trans. Magn.*, 2003, **39**(6), 3534–3540.
7. ETREMA Products Inc., *Terfenol-D Magnetostrictive Actuator Information*, Specifications, Public domain information, www.terfenoltruth.com, www.eterma.com.

7 Review of Giant Magnetoresistive (GMR) Materials

Mohsen Shahinpoor

Mechanical Engineering Dept., University of Maine, USA
Email: shah@maine.edu

7.1 Introduction

Magnetoresistance is defined as the property of a material to change its electrical conductivity or electrical resistivity if placed in an external magnetic field. In 1851, William Thomson (Lord Kelvin), discovered that when pieces of iron or nickel are placed within an external magnetic field, the electrical resistance increases when the current is in the same direction as the magnetic field, which is aligned with the magnetic N–S vector. He further noted that the resistance decreased when the current was perpendicular to the direction of the magnetic field,[1,2] and that the magnitude of changes in conductivity or resistivity was greater with nickel than iron. This magnetoresistance effect is referred to as anisotropic magnetoresistance (AMR). Lord Kelvin was unable to reduce the electrical resistance of any metal by more than about 5%. This effect is commonly known as the ordinary magnetoresistance (OMR) effect. There also exists the giant magnetoresistance (GMR) effect, discovered in 1988 independently by French and German scientists (Baibich and Fert *et al.*[3] and Binasch and Grünberg *et al.*[4]). Thus, we can differentiate

Fundamentals of Smart Materials
Edited by Mohsen Shahinpoor
© The Royal Society of Chemistry 2020
Published by the Royal Society of Chemistry, www.rsc.org

GMR from OMR. Note that GMR is also called colossal magnetoresistance (CMR), tunnel magnetoresistance (TMR) and extraordinary magnetoresistance (EMR). GMR materials generally possess alternating layers of ferromagnetic and non-magnetic but conductive layers such as iron–chromium and cobalt–copper. Magnetoresistance is defined as:

$$\delta H = [R(H) - R(0)]/R(0) \tag{7.1}$$

where $R(H)$ is the resistance of the sample in a magnetic field H, and $R(0)$ corresponds to initial resistance in the absence of a magnetic field. The term "giant magnetoresistance" indicates that the value δH for multilayer structures is much larger than ordinary magnetoresistance reduction by a few percent. Albert Fert of France and Peter Grünberg of Germany were jointly awarded the Nobel Prize in Physics in 2007, for the discovery of giant magnetoresistance and spintronics theory.[3,4]

Note that generally, the resistance of a conductor depends on the magnetization, which is controlled by the applied external magnetic field. The electrical resistance of conductors is due to collisions between conduction electrons and other charge carriers, electrons, ions or atoms. If a perfect metallic crystalline structure is at a temperature of absolute zero, it will allow an electron to move through it and experience no collisions so that the crystal would have zero resistance. Due to imperfections in the crystalline structure, at temperatures above absolute zero, the atoms vibrate out of their lattice locations and these imperfections and vibrations cause collisions, which increases the resistance of the metallic crystal. An applied external magnetic field can also increase the resistance of a material, since the magnetic force on the moving charges will tend to increase the number of collisions between charges. This dependence of resistance on the magnetic field is called the **magnetoresistivity effect**. Note that changes in resistance are directly proportional and related to the strength of the applied external magnetic field. The magnetoresistivity effect is employed in computers to read magnetic data off CDs and magnetic storage electronic media such as magnetic tapes. Normally, a potential difference is applied to an electronic circuit, which is placed close to the magnetic material on disk or tape. As the dynamic magnetic fields affect the wire, the resistance of the wire changes with the magnetic field. This change in resistance changes the current through the wire and provides a reading of the magnetic field on the tape or disk. Magnetoresistance is capable of providing more accurate reading data than magnetic induction.

Magnetoresistance behavior of materials depends on the imposed magnetic field rather than the time variations of the magnetic field. Measurements of the effect are obtained by dividing the change in resistance (or change in current) by the value of the magnetic field of the storage medium. Note that inductive heads can give about a 1% effect, while magnetoresistance heads give about a 4% effect.[5]

Exercise No. 7.1 In how many ways does the magnetoresistive effect manifests itself?

Answer to Exercise No. 7.1 The basic effect is called the ordinary magnetoresistance (OMR) effect. There are also the giant magnetoresistance (GMR), colossal magnetoresistance (CMR), tunnel magnetoresistance (TMR) and extraordinary magnetoresistance (EMR).

Exercise No. 7.2 What distinguishes the GMR from the OMR?

Answer to Exercise No. 7.2 GMR materials generally possess alternating layers of ferromagnetic and non-magnetic but conductive layers such as iron–chromium, and cobalt–copper have been tested. Besides this, the OMR changes the resistance by as much as 5% or less, but GMR induce changes in resistivity (~100%) of significantly higher than 5%.

7.2 Ordinary Magnetoresistance (OMR)

To further analyze the phenomenon of magnetoresistance, let us briefly describe the nature of electromagnetic forces at work.[6,7] When charged entities such as electrons or ions move in a magnetic field, they experience a magnetic force given by:

$$\mathbf{F}_B = q\mathbf{v} \times \mathbf{B}, \tag{7.2}$$

where the boldface on $\mathbf{F}$, $\mathbf{v}$, and $\mathbf{B}$ indicates that these quantities are vectors and $\times$ represents a vector cross product operator. The scalar magnitude of the force is given by eqn (7.3):

$$|\mathbf{F}_B| = q\,|\mathbf{v}|\,|\mathbf{B}|\,\sin\,\theta, \tag{7.3}$$

with θ representing the angle between $\mathbf{B}$ and $\mathbf{v}$. The resulting magnetic force, $\mathbf{F}_B$, acts on a charge moving through a conductive wire. If the charge of the moving charges is negative, $q\mathbf{v}$ points opposite the direction of $\mathbf{v}$ such that the wire is in a uniform magnetic field. The magnetic force developed disrupts the normal flow of electrons, causing more collisions with atoms and other electrons, thus increasing the resistance of the wire. Based on quantum mechanical principles,[3,4] if conduction electrons moving through a magnetoresistive medium have inherent spins in the same direction as the imposed magnetic field, they experience fewer collisions with other entities (electrons, ions, atoms, *etc.*) than electrons with inherent spins in the opposite direction of the imposed magnetic field. If, however, the magnetoresistive medium is Fe–Cr layered, the resistance can change dramatically, leading to the phenomenon of the **giant magnetoresistance effect.** Giant magnetoresistance did get its name by showing that the applied magnetic field can give ratios of over 100% for electrical conductivity or resistance. Note that regular magnetoresistance gives rise to around a 5% effect on resistance and about 1% effect on inductance. IBM produced the first GMR hard drive in 1997, and the use of the new technology has gradually increased. In the next section, GMR is further expanded upon and discussed.

Exercise No. 7.3 What % electrical resistances in magnetoresistive materials can change?

Answer to Exercise No. 7.3 Regular magnetoresistance (OMR) gives rise to about a 5% effect on resistance, while giant magnetoresistance shows that the applied magnetic field can change the local resistance by over 100%.

7.3   Spintronics and GMR Effect

GMR materials were developed independently by Peter Grünberg and Albert Fert in 1988.[3,4,8,9] Grünberg's team tested thin chromium films, of approximately one nanometer thick, between thicker iron films (about 8 nm), producing antiferromagnetic coupling (Figure 7.1). Magnetization orientation of each iron film changes direction to produce coupling across the chromium film (Figures 7.1 and 7.2). According to Grünberg, the resistance changes were due to electron scattering in the chromium layer. Grünberg's team tested strips of a double layer of iron films, and also tested triple-layered iron films. The triple-layered sample exhibited approximately twice the response of the double layered sample.[3,4]

Note that GMR is a completely different phenomenon from magnetoresistance (MR). MR utilizes the regular "Lorenz" force on electrons moving in a magnetic field. GMR exploits a new phenomenon called "Spintronics", which exploits the spin-dependent scattering of electrons.

Fert *et al.*[3,8] reported resistance reduced by a factor of nearly two when the saturation magnetization was reached at temperatures of 4.2 K. The "superlattices" they produced were also iron–chromium

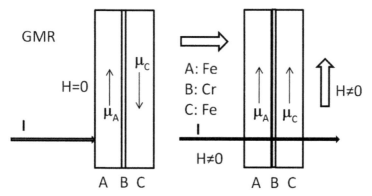

Figure 7.1 Magnetization directions in Fe–Cr layers showing how conductivity increases upon aligning in the same direction the magnetization vectors of ferromagnetic layers. When magnetization vectors are not aligned, the electrons suffer from scattering at the middle layer due to spintronics effects, as shown in work by Fert *et al.*[3,8] and Grünberg *et al.*[4,9]

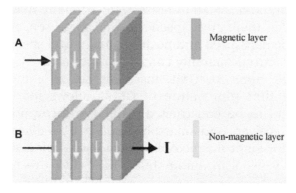

Figure 7.2 Alignment of magnetic orientations establishes the GMR effects for current perpendicular to the plane (CPP).

layers, with thicknesses of 9 Å for the chromium layers and 30 Å for the iron layers. At room temperature, the magnetoresistance at saturation reduced by a factor of two, but was still significant. Fert *et al.* also suggested that electron scattering was occurring across the chromium layers due to the antiparallel spin orientations of the iron layers.[3]

While spin scattering had been suggested independently by Grünberg *et al.*[4,9] and Fert *et al.*,[3,8] the first theoretical modeling of the mechanism was undertaken by Camley and Barnaś.[10] They were able to reproduce the behavior of superlattices in their model, although they noted that the model did not account for anisotropic magnetoresistance, which might be important in ferromagnets other than iron.[10]

GMR materials are used in sensor applications, particularly as read heads in hard disk drives. In the read head, changing magnetic fields encoding the bits drive resistance changes in the magnetoresistive sensor. The particular device in this sensor is termed a spin valve. Spin valves incorporate an antiferromagnetic layer to pin the magnetization direction of one layer in the superlattice. The other magnetic layer can reorient to an applied magnetic field, resulting in a high sensitivity to external magnetic fields (Baraduc *et al.*[11]).

GMR sensors have been used in the automotive industry to replace Hall and AMR sensors. Broadly categorized into speed-sensing and angle-sensing applications, giant magnetoresistivity offers a high sensitivity without much noise. As angle sensors, they are used to control the commutation of electric motors. They are also used to measure the angle of the steering wheel for electronic stability control, lane departure warnings, and active suspension. The inherent high sensitivity and signal-to-noise ratio of giant magnetoresistive materials makes them great speed sensors. Speed sensors are used to monitor crankshaft speed and position, and wheel speed for antilock brakes and electronic stability control (Kasper *et al.*[12]).

Other applications for GMR materials are as compass magnetometers. The thin film nature of GMRs allows for very compact magnetometers to be constructed. They are incorporated in commercial GPS receivers and smartphones. Since its discovery in 1988, applications for giant magnetoresistance have exploded. The utilization of GMR sensors in automobiles, hard disk drives, and smartphones means that many aspects of modern life depend on our understanding of GMR, and our ability to improve the response of sensors utilizing it.

Exercise No. 7.4 What are spin valves?

Answer to Exercise No. 7.4 The particular device used in magnetoresistive sensors is called a spin valve. Spin valves incorporate an antiferromagnetic layer to pin the magnetization orientation of one layer in the superlattice.

7.4 Applications of GMR

Since its discovery in 1988, many applications for GMR have emerged and been realized. For example, GMR sensors have been extensively used in automobiles, hard disk drives, and smartphones. GMR materials are used extensively as reading heads in hard disk drives. The particular device in this sensor is termed a spin valve. Spin valves incorporate an antiferromagnetic layer to pin the magnetization orientation of one layer in the superlattice. The other magnetic layer is thus capable of reorienting to an applied magnetic field, resulting in a high sensitivity to external magnetic fields (Baraduc *et al.*[11]).

GMR sensors are broadly used in speed-sensing and angle-sensing applications. They offer a high sensitivity without much noise. As angle sensors, they are used to control the commutation of electric motors. They are also used to measure the angle of the steering wheel for electronic stability control, lane departure warnings, and active suspension. As emphasized before, speed sensors are particularly dependent on the high sensitivity, and the high signal-to-noise ratio that GMRs can offer. Speed sensors are used to monitor crankshaft speed and position, and wheel speed for antilock brakes and electronic stability control (Kasper *et al.*[12]). The thin film nature of GMRs allows for very compact magnetometers to be constructed. They are incorporated in commercial GPS receivers and smartphones.

7.5 Modeling

According to Coehoorn,[15,20] the total conductivity of magnetoresistive materials can be represented as the sum of contributions from spin-up and spin-down electrons. Since the spin quantum number of the conduction electrons is conserved in the scattering processes, at least at low temperatures, the total conductivity depends on spin-up and spin-down electrons.

For magnetic-nonmagnetic layered (Co–Cu) structures in the presence of parallel magnetizations, the GMR effect arises when the layer-averaged electron mean-free path is larger than the multilayer period, as well as being different from that for the other spin direction. This model is depicted schematically in Figure 7.3 for the Co/Cu system. Figure 7.3 only shows electron scattering at the interfaces of cobalt and copper. Scattering is strong for electrons with antiparallel spin to the magnetization direction of the cobalt layers, and weak for electrons with parallel spin.[20] Note that each layer acts like a spin valve in the sense that its magnetization direction determines whether it allows spin-up or spin-down electrons to pass through or get scattered back and not go through the layers. In the case of parallel-aligned magnetic layers, the resistivity for one spin channel is low, leading to the low total resistance. However, antiparallel alignment of alternate

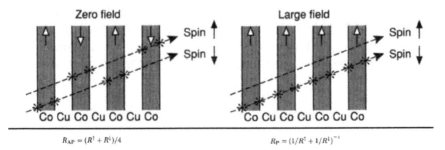

Figure 7.3   Parallel and antiparallel cases for GMR.

magnetic layers results in appreciable scattering for electrons in both spin directions, and hence in a larger resistance.

Supposing that $R\uparrow$ and $R\downarrow$ are the material resistances for the spin-up and spin-down electrons, respectively, thus, in the parallel-layers case, the resistances for the parallel (P) and anti-parallel (AP) cases are denoted by R_P and R_{AP}, respectively, such that:

$$R_{AP} = (1/4)(R\uparrow + R\downarrow) \qquad (7.4)$$

$$R_P = (1/4)(1/R\uparrow + 1/R\downarrow)^{-1} \qquad (7.5)$$

and are shown in Figure 7.3 for the cobalt–copper system.

Defining the scattering asymmetry parameter as $\alpha = R\uparrow/R\downarrow$, the magnetoresistance ratio (MR) can be expressed as:

$$(R_{AP} - R_P)/R_P = (1 - \alpha)^2/4\alpha \qquad (7.6)$$

$R_{AP} > R_P$, and the MR ratio is positive. It should be noted that the MR ratio can either be positive or negative for systems containing different ferromagnetic layers.[3,4]

7.6   Role of Electron Spin in GMR

Electrons possess intrinsic magnetic fields due to their spin. In the absence of magnetization of a ferromagnetic material, half of the conduction electrons possess a positive spin (spin-up) along the chosen axis, and half will have a negative spin (spin-down).

If a ferromagnetic material is magnetized along the positive spin direction, a spin-up electron will travel through the material easier than a spin-down electron.[3,4,8,9]

7.7 GMR in Granular Structures

Rubinstein *et al.*[13] and Granovsky *et al.*[14] reported a GMR effect in granular mixtures or powders of ferromagnetic and non-magnetic metals. Here, the spin-dependent scattering of electrons occurs at the surface and in the bulk of the grains.[13,14]

According to,[13,14] the granular mixture of ferromagnetic and non-ferromagnetic metals form powder-like ferromagnetic nanoclusters of around 10 nm in size within particles of 10 nm in diameter embedded in a non-magnetic metallic powder, forming a superlattice.

7.8 GMRs as Smart Sensors

GMRs can be used as smart sensors to detect the presence and variations of magnetic fields. For example, in hard disk drives as well as biosensors[15,16] and as detectors of oscillations in microelectromechanical systems (MEMS) and nanoelectromechanical systems (NEMS). GMR sensors are multi-layered consisting of a silicon substrate, a binder layer, sensing layer, non-magnetic layer, fixing or pinning layer, antiferromagnetic layer and protective layer.[15–19] A typical non-magnetic material used is copper, and the binder and protective layers are formed by metals such as tantalum. The pinning layer, which is made of a magnetic material such as cobalt establishes the direction of its magnetization.

7.9 Hard Disk Drives[19]

Hard drives work based on the principles of magnetic domain variations such that 1 is associated with a change in the direction of their magnetization, while no change represents a logical 0, and this information can be digitally transferred back and forth. The direction of the magnetic field with the domain walls and the magnetization direction is normally perpendicular to the surface in the antiferromagnetic conductive layer and parallel to the surface in the ferromagnetic sensing layer.

7.10 Conclusions

This chapter presented a brief review of the GMR effect originally and independently discovered and developed by Peter Grünberg and

Albert Fert in 1988. Thin chromium films, approximately one nanometer thick, between thicker iron films (around 8 nm), can produce antiferromagnetic coupling. The magnetization orientation of each iron film changes direction to produce the coupling across the chromium film. The resistance changes were due to electron scattering in the chromium layer. The triple-layered sample exhibited approximately twice the response that of the double layered sample. Note that GMR is a completely different phenomenon than MR. MR utilizes the regular "Lorenz" force on electrons moving in a magnetic field. GMR exploits a new phenomenon called "Spintronics", which exploits the spin-dependent scattering of electrons, producing a GMR effect.

Homework Problems

Homework Problem 7.1

Describe the concepts of spin-up and spin-down electrons.

Homework Problem 7.2

How can traveling electrons in ferromagnetic materials experience fewer or more collisions with other electrons and atoms?

Homework Problem 7.3

How are the electrons scattered in GMR materials?

Homework Problem 7.4

Present a simple model for GMR effects.

References

1. D. Mapps, Magnetoresistive sensors, *Sens. Actuators, A*, 1997, **59**(1), 9–19.
2. C. Reig, S. Cardoso and S. Mukhopadhyay, *Giant Magnetoresistance (GMR) Sensors: From Basis to State-of-the-art Applications*, Springer, London, New York, 2013.
3. M. N. Baibich, J. M. Broto, A. Fert, F. Nguyen Van Dau, F. Petroff, P. Etienne, G. Creuzet, A. Friederich and J. Chazelas, Giant Magnetoresistance of (001) Fe/(001) Cr Magnetic Superlattices, *Phys. Rev. Lett.*, 1988, **61**(21), 2472–2475.
4. G. Binasch, P. Grünberg, F. Saurenbach and W. Zinn, Enhanced Magnetoresistance in Layered Magnetic Structures with Antiferromagnetic Interlayer Exchange, *Phys. Rev. B*, 1989, **39**(7), 4828–4830.

5. L. Jogschies, D. Klaas, R. Kruppe, J. Rittinger, P. Taptimthong, A. Wienecke, L. Rissing and M. C. Wurz, Review Recent Developments of Magnetoresistive Sensors for Industrial Applications, *Sensors*, 2015, **15**, 28665–28689.
6. J. D. Livingston, *The Natural Magic of Magnets*, Harvard University Press, Cambridge, Massachusetts, 1996.
7. T. R. McGuire and R. I. Potter, Anisotropic Magnetoresistance in Ferromagnetic 3d Alloys, *IEEE Trans. Magn.*, 1975, **11**(4), 1018–1037.
8. C. Chappert, A. Fert and F. Nguyen Van Dau, The emergence of spin electronics in data storage, *Nat. Mater.*, 2007, **6**(11), 813–823.
9. P. Grünberg, R. Schreiber, Y. Pang, M. B. Brodsky and H. Sowers, Layered Magnetic Structures: Evidence for Antiferromagnetic Coupling of Fe Layers across Cr Interlayers, *Phys. Rev. Lett.*, 1986, **57**(19), 2442–2445.
10. R. E. Camley and J. Barnaś, Theory of giant magnetoresistance effects in layered magnetic structures with antiferromagnetic coupling, *Phys. Rev. Lett.*, 1989, **63**(6), 664–667.
11. C. Baraduc, M. Chshiev and B. Dieny, Spintronic Phenomena: Giant Magnetoresistance, Tunnel Magnetoresistance, and Spin Transfer Torque, in *Giant Magnetoresistance (GMR) Sensors: From Basis to State-of-the-Art Applications*, ed. C. Reig, S. Cardoso de Freitas and S. C. Mukhopadhyay, Springer, New York, 2013, pp. 1–30.
12. K. Kasper, M. Weinberger, W. Granig and P. Slama, GMR Sensors in Automotive Applications, in *Giant Magnetoresistance (GMR) Sensors: From Basis to State-of-the-art Applications*, ed. C. Reig, S. Cardoso de Freitas and S. C. Mukhopadhyay, Springer, New York, 2013, pp. 133–156.
13. M. Rubinstein, B. N. Das, N. C. Koon, D. B. Chrisey and J. Horwitz, Ferromagnetic-resonance studies of granular giant-magnetoresistive materials, *Phys. Rev. Part B*, 1994, **50**(1), 184–192.
14. A. B. Granovsky, M. Ilyn, A. Zhukov, V. Zhukova and J. Gonzalez, Giant magnetoresistance of granular microwires: Spin-dependent scattering in intergranular spacers, *Phys. Solid State*, 2011, **53**(2), 320–322.
15. R. Coehoorn, Novel Magnetoelectronic Materials and Devices. Giant magnetoresistance and magnetic interactions in exchange-biased spin-valves, Lecture Notes. Technische Universiteit Eindhoven. Archived from the original (PDF) on 2011-08-10. Retrieved 2011-04-25, 2003.
16. D. L. Graham, H. A. Ferreira and P. P. Freitas, Magnetoresistive-based biosensors and biochips, *Trends Biotechnol.*, 2004, **22**(9), 455–462.
17. E. Hirota, H. Sakakima and K. Inomata, *Giant Magneto-Resistance Devices*, Springer, London, New York, 2002.
18. D. Rifai, A. N. Abdalla, K. Ali and R. Razali, Giant Magnetoresistance Sensors: A Review on Structures and Non-Destructive Eddy Current Testing Applications, *Sensors*, 2016, **16**(3), 298–304.
19. E. Hill, *et al.*, A giant magnetoresistive magnetometer, *Sens. Actuators, A*, 1997, **59**(1), 30–37.
20. R. Coehoorn, Giant Magnetoresistance in Exchange-Biased Spin-Valve Layered Structure and its Application in Read Heads, *Springer Series in Surface Sciences*, Springer Publications, London/New York, 2000.

8 Review of Magnetic Gels as Smart Materials

Mohsen Shahinpoor

Mechanical Engineering Dept., University of Maine, USA
Email: shah@maine.edu

8.1 Introduction

Magnetic gels belong to the family of hydrogels, polymeric gels and are generally polyelectrolyte gels. They are highly swollen molecular networks, which are crosslinked to create a hydrophilic solid. Kuhn, Kunzle, and Kathchalsky,[1] Kuhn,[2] Katchalsky,[3] Kuhn, Hargitay, Katchalsky, and Eisenberg[4] were the first to report reversible swelling and deswelling of polymeric gels by ionization. Later, Tanaka, Nishio, Sun, and Ueno-Nishio[5] showed that such gels can collapse in an electric field. Later, Osada,[6] De Rossi, Kawana, Osada and Yamauchi,[7] De Rossi, Parini, Chiarelli and Buzzigoli,[7] Osada[8] and Schneider, Kato and Strongin[9] showed that such swelling and deswelling can also be induced chemically (pH Muscles) or electrically (EAP).

Zrinyi and co-workers[10–19] were the first to develop magnetically active gels with responses that can be vastly accelerated using an applied magnetic field. This chapter is a compact review of magnetic gels based on related research and development performed by Zrinyi and co-workers.[10–19]

Note that magnetic gels are considered to be members of the smart materials family and in ways are similar to soft silicon rubber magnetic composites, used in our daily life as soft magnetic stickers.

Fundamentals of Smart Materials
Edited by Mohsen Shahinpoor
© The Royal Society of Chemistry 2020
Published by the Royal Society of Chemistry, www.rsc.org

However, magnetic gels are softer and more stretchable and maneuverable in a magnetic field and can sustain soft actuation at the micro and nano levels. A prelude to the development of ferrogels was a classic paper by Rosenzweig[20] published in 1985 on ferrohydrodynamics. Ferrogels are chemically crosslinked polymer networks swollen by a colloidal ferrofluid. Ferrofluids are also called magnetic fluids and are colloidal dispersions of monodomain magnetic particles.[20] Typically, the single domain colloidal magnetic particles have dimensions of around 10–15 nm. Furthermore, they are superparamagnetic, in which magnetization can randomly flip direction under the influence of temperature.[21]

Magnetic gels or ferrogels belong to the general family of magnetostrictive materials, which produce strain when exposed to a magnetic field. One may also embed magnetic coils within these materials to be able to also electrically control the deformation of ferrogels.

As described above, single domain, magnetic particles of colloidal size can be incorporated into chemically crosslinked polyvinyl alcohol-type hydrogels to make a colloidal magnetic suspension or ferrogel. Ferrogels undergo quick and reversible shape transformation brought about by changes in their external magnetic field. Elongation, bending, contraction, and expansion can be observed in magnetic gels by proper orientation of the external magnetic field. Figures 8.1 and 8.2 represent some examples of their applications

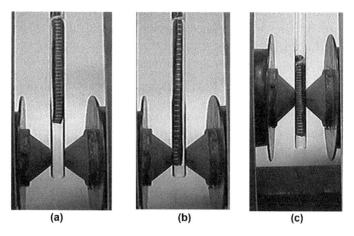

(a)   (b)   (c)

Figure 8.1 Shape distortion of ferrogels due to a non-uniform magnetic field, (a) no external magnetic field, (b) the maximal field strength is located under the lower end of the gel, and (c) the maximal field strength is focused in the middle of the gel along its axis. Reproduced from ref. 19 with permission from the Royal Society of Chemistry.

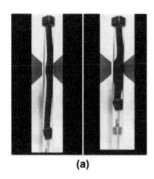 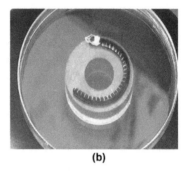

(a) **(b)**

Figure 8.2 (a) Elongation and contraction in magnetic fields as artificial muscles and (b) bending induced by a permanent magnet. Reproduced from ref. 19 with permission from the Royal Society of Chemistry.

such as the slithering of an artificial snake, crawling and walking on a constraint line, behaving like biomimetic artificial muscles, plus many more interesting dynamic applications.

8.2 Magnetoviscoelasticity of Ferrogels

Ferrogels consist of fine, distributed magnetic particles in a swelling liquid such as polyacrylamide hydrogels. These magnetic particles (Fe, iron) are attached to the flexible molecular network chains of the swelling liquid by adhesive forces. The solid particles carry a magnetic moment, m. In this way, in the absence of an applied external magnetic field, the magnetic moments are oriented in a random manner, creating a magnetic gel with no net magnetization.

When an external magnetic field is applied to a ferrogel, the internal magnetic moment on each colloidal ferromagnetic particle experiences an imposed magnetic moment, which aligns the ferromagnetic particles with the direction of the external field. This alignment produces a bulk magnetic moment in the ferrogel. Magnetization saturation will set in upon increasing the strength of the external magnetic field, which aligns all of the particle magnetic moments along the field lines. If the external magnetic field is now turned off, the magnetic dipole moments lose their magnetic moment and randomize quickly to reduce the bulk magnetization to nearly zero. Ferrogels behave very similar to hydrogels with non-magnetic colloidal particles if the magnetic field is absent.

Generally speaking, a magnetic gel behaves like a viscoelastic solid and does react to temperature variations. The externally imposed magnetic field has to be spatially nonuniform to generate any magnetic force on a ferrogel sample. Thus, the spatial variations of the external magnetic field generate a bulk magnetic moment on the ferrogel sample. When the external magnetic field is uniform with no spatial variations, a ferrogel experiences no net force in such uniform fields. On the other hand, in a non-uniform magnetic field, magnetic moments and forces act on the colloidal magnetic particles and move them around within the network, and that movement causes the network to deform. Ferrogels in a magnetic field are deformed, and the deformation can be controlled, and the final shape is established by the balance of the magnetic and viscoelastic interactions. Magnetic gels in a spatially non-uniform magnetic field experience magnetic forces acting on the magnetic particles. Ferromagnetic particles embedded within the gel tend to move due to the imposed spatially non-uniform magnetic field. The externally applied spatially non-uniform magnetic field generates a displacement such that the final shape is set by the balance of magnetic and elastic interactions within the ferrogel. According to Zrinyi *et al.*,[10-19] the magnetic force density, $\mathbf{f}_m$, of a ferrogel can be written as:

$$\mathbf{f}_m = \mu_0 \, (\mathbf{M}\nabla) \, \mathbf{H} \tag{8.1}$$

where μ_0 is the magnetic permeability of the vacuum, $\mathbf{M}$ represents the magnetization matrix, and the $\nabla\mathbf{H}$ matrix takes into account the gradient of the imposed spatially non-uniform magnetic field, $\mathbf{H}$. Note that the force vector, $\mathbf{f}_m$, is parallel to the direction of the applied magnetic field.

8.3 Constitutive Equations for Ferrogels

Following the work by Zrinyi *et al.*,[10-19] one may assume magnetic gels or ferrogels behave similarly to hyperelastic materials and in particular to Neo-Hookean solids (rubber elasticity) so that one can derive expressions for various stress components in hydrogels to derive their governing dynamic equilibrium equations.

Note that the constitutive equations for a hyperelastic material can be derived by relating the free energy of the material to the deformation gradients and in particular to the three strain invariants I_1, I_2 and I_3 of the strain tensor E_{ij} ($i, j = 1, 2, 3$) such that these invariants are given by eqn (8.2)–(8.4).

$$I_1 = E_{11} + E_{22} + E_{33} = \text{Trace } E_{ij} = E_{kk} \tag{8.2}$$

$$I_2 = (1/2)(I_1^2 - E_{ik}E_{ki}) \tag{8.3}$$

$$I_3 = \det E_{ij} = J^2 \tag{8.4}$$

The Lagrange strain tensor E_{ij} is defined as:[22]

$$E_{kk} = (1/2)\left\{ \frac{\partial u_i}{\partial x_j} + \frac{\partial u_j}{\partial x_i} + \frac{\partial u_k}{\partial x_j}\frac{\partial u_k}{\partial x_j} \right\} \tag{8.5}$$

where u_i, $(i = 1, 2, 3)$ is the deformation vector and x_i, $(i = 1, 2, 3)$ are the spatial coordinates for the magnetic gel with X as the material coordinate, as shown in Figure 8.3.

Let e_1, e_2, and e_3 denote the three eigenvalues of E_{ij}. The principal stretches are then given by:

$$\lambda_1 = \sqrt{e_1}, \lambda_2 = \sqrt{e_2}, \lambda_3 = \sqrt{e_3} \tag{8.6}$$

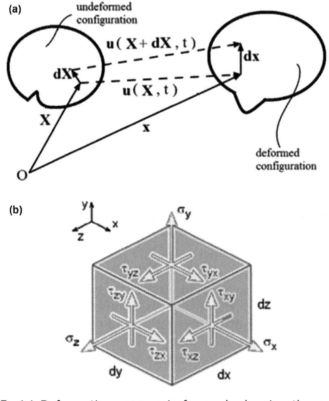

Figure 8.3 (a) Deformation process in ferrogels showing the undeformed and deformed states of a differential element with x as the spatial and X as the material coordinates and (b) the three-dimensional state of stress.

Note that the three principal stretches are related to the invariants of strain I_1, I_2 and I_3 such that:

$$I_1 = \lambda_1^2 + \lambda_2^2 + \lambda_3^2 \tag{8.7}$$

$$I_2 = \lambda_1^2\lambda_2^2 + \lambda_2^2\lambda_3^2 + \lambda_3^2\lambda_1^2 \tag{8.8}$$

$$I_3 = \lambda_1^2\lambda_2^2\lambda_3^2 \tag{8.9}$$

The stress–strain law can now be deduced by differentiating the free energy U with respect to the strain invariants **I1, I2**, and **I3**, such that:[22]

$$\sigma_{ij} = (2/J)\left\{ \left[\frac{\partial U}{\partial I_1} + I_1 \frac{\partial U}{\partial I_2}\right]E_{ij} - \frac{\partial U}{\partial I_2}E_{ik}E_{kj} \right\} + 2J\left(\frac{\partial U}{\partial I_3}\right)\delta_{ij} \tag{8.10}$$

where, δ_{ij} is the Kronecker delta function or 3×3 identity matrix.

The hydrostatic stress p is an unknown variable, which must be calculated by solving the associated boundary value problem. Note that for a generalized Neo-Hookean solid[19,22] the strain energy reduces to:

$$U = (G/2)(I_1 J^{2/3} - 3) + (E/2(1 - 2\nu))(J - 1)^2 \tag{8.11}$$

where G and E are the shear modulus and Young's modulus of the magnetic gel, respectively, and ν is the Poisson's ratio of the magnetic gel. Note that $G = E/2(1 + \nu)$. For nearly incompressible Neo-Hookean gels, the strain energy takes a simple form, as shown in eqn (8.12):

$$U = (G/2)(I_1 - 3) = (G/2)\,(\lambda_1^2 + \lambda_2^2 + \lambda_3^2 - 3) \tag{8.12}$$

Note that in this case due to incompressibility, $J = \lambda_1\lambda_2\lambda_3 = 1$, and thus, for a uniform bar of ferrogel, $\lambda_2 = \lambda_3 = \lambda_1^{(-1/2)}$ and thus:

$$U = (G/2)(\lambda_1^2 + \lambda_2^2 + \lambda_3^2 - 3) = (G/2)\,(\lambda_1^2 + 2\lambda_1^{-1} - 3) \tag{8.13}$$

Thus, it can be deduced from eqn (8.11) that:

$$\sigma_{11} - \sigma_{33} = \lambda_1 \frac{\partial U}{\partial \lambda_1} - \lambda_3 \frac{\partial U}{\partial \lambda_3}, \quad \sigma_{22} - \sigma_{33} = \lambda_2 \frac{\partial U}{\partial \lambda_2} - \lambda_3 \frac{\partial U}{\partial \lambda_3} \tag{8.14}$$

Since in this case, $\sigma_{22} = \sigma_{33} = 0$, the expression for uniaxial actual stress, σ_{11}, can be derived such that:

$$\sigma_{11} = \lambda_1 \frac{\partial U}{\partial \lambda_1} = \lambda_1 G(\lambda_1 - \lambda_1^{-2}) = G(\lambda_1^2 - \lambda_1^{-1}) \tag{8.15}$$

Note that σ_{11} is the true stress or axial force applied to the ferrogel rod divided by the deformed cross-section of the rod, which in this case is related to the initial undeformed cross section by the product of λ_2 and λ_3, which is equal to $\lambda_2\lambda_3 = \lambda_1^{-1}$. Thus, the uniaxial engineering stress $\sigma_{11}^{eng} = \sigma_n = \lambda_1^{-1}\sigma_{11}$ in the ferrogels can be expressed as:[19]

$$\sigma_n = G(\lambda_z - \lambda_z^{-2}) \tag{8.16}$$

Here, the nominal stress, σ_n, is defined as the ratio of the equilibrium elastic force to the undeformed cross-sectional area of the sample, and λ_z is the stretch in the z-direction.

Exercise No. 8.1: Derive eqn (8.16) from the generalized constitutive equations for rubber elasticity.

The solution to Exercise No. 8.1: From eqn (8.15) note that:

$$\sigma_{11} = \lambda_1(\partial U/\partial\lambda_1) = \lambda_1 G(\lambda_1 - \lambda_1^{-2}) = G(\lambda_1^2 - \lambda_1^{-1}) \tag{8.17}$$

and

$$\sigma_{11}^{eng} = \sigma_n = \lambda_1^{-1}\sigma_{11}. \tag{8.18}$$

$$\sigma_{11}^{eng} = \sigma_n = \lambda_1^{-1}\sigma_{11} = \lambda_1^{-1}G(\lambda_1^2 - \lambda_1^{-1}) = G(\lambda_1 - \lambda_1^{-2}) = G(\lambda_z - \lambda_z^{-2}) \tag{8.19}$$

8.4 Analysis of Dynamics of Magnetic Gel Actuators in a Magnetic Field

Let us consider a vertically suspended cylindrical magnetic gel cylindrical strand intended to pull a weight upward. Initially, due to the weight of the strand, the gel is preloaded with its weight $M^g g$, where M^g is the mass of the magnetic gel strand.

Thus, the gel is under a passive stretch, λ_m, due to its weight. The nominal stress due to this passive stretch can be written as:

$$\sigma_n = \frac{M^g g}{a_0} \tag{8.20}$$

where M^g is the mass of the magnetic gel, g denotes the gravitational acceleration and a_0 denotes the initial undeformed cross-section of the magnetic gel strand. Let us rewrite eqn (8.17) in the form of a cubic equation to solve for λ_M:

$$\lambda_M^3 - \alpha\lambda_M^2 - 1 = 0 \tag{8.21}$$

$$\alpha = \frac{\sigma_n}{G} = \frac{M^g g}{a_0 G} \tag{8.22}$$

Here a_0 denotes the undeformed cross-sectional area of the gel at rest. Eqn (8.21) shows us how the passive stretch, λ_M, depends on the applied mechanical stress and the shear modulus, G.

Exercise No. 8.2 Employing eqn (8.21) and (8.22), show that the passive stretch $\lambda_M \geq 1$

The solution to Exercise No. 8.2 Note that from eqn (8.21):

$$\lambda_M^3 = \alpha \lambda_M^2 + 1 \geq 1 \tag{8.23}$$

Thus, in the presence of a load or stress, the magnetic gel tends to elongate, *i.e.*, $(\lambda_M > 1)$.

Under an applied magnetic field, the magnetic gel strand will develop a new total stretch, $\lambda_{M,H}$, as well as magnetization, **M**. The magnetization, **M**, of a magnetic gel is defined by the Langevin function based on the assumption that the magnetization of individual colloidal particles in the magnetic gel is approximately equal to the saturation magnetization of the pure ferromagnetic material. Thus, the magnetization, M, of magnetic gels, in an applied magnetic field can be expressed as:[19,20]

$$M = \Phi_m M_s L(\xi) = \Phi_m M_s \left(co\,th\,\xi - \frac{1}{\xi} \right) \tag{8.24}$$

where Φ_m stands for the volume fraction of the magnetic particles in the whole gel, and ξ of the Langevin function $L(\xi)$ is defined as

$$\xi = \frac{\mu_0 m H}{k_B T} \tag{8.25}$$

In eqn (8.25), m stands for the magnetic moment of embedded particles, k_B is the Boltzmann constant and T stands for the ferrogel temperature. Eqn (8.24) establishes that the magnetization of a ferrogel is proportional to the volume fraction of embedded magnetic particles and their saturation magnetization. Assuming a linear relationship between the magnetization, m, and magnetic field strength, H, and homogeneous deformation for he ferrogel sample, one obtains a linear relationship between the magnetization, M, and magnetic field strength, H, such that this latter assumption is valid at small magnetic field intensities. In this case, a Taylor series expansion of the Langevin distribution function in eqn (8.24) yields: $L(\xi) \cong \xi/3$, according to work by Zrinyi.[19] This results in a linear relationship between the magnetization and field intensity:

$$M = \chi H, \quad H \to 0 \tag{8.26}$$

where χ denotes the initial susceptibility, which is defined as

$$\chi = \Phi_m M_S \frac{m}{3k_B T} \tag{8.27}$$

Note that according to Zrinyi,[19] for ferrogels $M_S m/3k_B T = 0.338$. The magnetic force can be determined using eqn (8.1). If we assume the total stress to be mechanical stress plus magnetic stress, then one can derive the following equations:[19]

$$\lambda_{M,H}^3 - \alpha \lambda_{M,H}^2 - \beta(H_h^2 - H_m^2)\lambda_{M,H} - 1 = 0 \tag{8.28}$$

where H_h and H_m are the magnetic field strengths at the bottom and the top of a rod-like ferrogel fiber, respectively (see Figure 8.4).

The parameter β can be considered as the stimulation coefficient, defined as:

$$\beta = \frac{\mu_0 \chi}{2G} \tag{8.29}$$

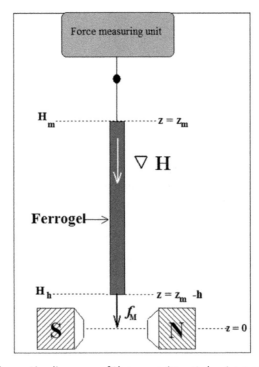

Figure 8.4   Schematic diagram of the experimental set-up used to study the magnetoelastic properties of the cylindrical form of ferrogels. Reproduced from ref. 19 with permission from the Royal Society of Chemistry.

Eqn (8.28) establishes that elongation occurs, *i.e.*, $\lambda_{M,H} > \lambda_M > 1$, if we suspend a ferrogel fiber in a nonhomogeneous magnetic field in such a way that $H_h > H_m$. On the other hand, when the field is turned off, the stress is lifted. In the opposite case, when $H_h < H_m$ work is released when the magnetic field is applied, *i.e.*, $\lambda_M > 1$, but $\lambda_{M,H} < \lambda_M$. The displacement of the load can be expressed as:

$$\Delta h = h_0(\lambda_{M,H} - \lambda_M), \tag{8.30}$$

The displacement of the end of the gel is found to be much more significant at elongation than at contraction.[19] An expression for the mechanical work is given by eqn (8.31):

$$W = -M^g g \Delta h = -M^g g h_0 \left(\lambda_{M,H} - \lambda_M\right) \tag{8.31}$$

where h_0 denotes the undeformed length of the magnetic gel fiber. Note that the condition for uniaxial deformation of a ferrogel cylinder can be written as follows:[10–19]

$$\lambda_H^3 - \beta(H_h^2 - H_m^2)\lambda_H - 1 = 0 \tag{8.32}$$

where λ_H denotes the ratio of the deformation stretch due to the magnetic field induced strain. The parameter β is given by eqn (8.33):

$$\beta = \frac{\mu_0 \chi}{2G} \tag{8.33}$$

where χ stands for the initial susceptibility of the ferrogel, and H_h and H_m represent the magnetic field strengths at the bottom and the top of a suspended ferrogel cylinder, respectively.

According to Zrinyi,[19] the z-directional distribution of the magnetic field strength can be satisfactorily approximated by the following forms:

$$h(z) = \begin{cases} 1 - kz^2 & \text{if} \quad |z| < \delta \\ (1 - k\delta^2)e^{-\gamma(|z|-\delta)} & \text{if} \quad |z| \geq \delta \end{cases} \tag{8.34}$$

where γ is a characteristic constant describing the exponential decay of field strength at larger distances, δ represents the radius of poles, and the constant k is given by:

$$k = \frac{\gamma}{2\delta + \gamma\delta^2} \tag{8.35}$$

This discussion completes our initial analysis of the dynamics of ferrogel actuators in a magnetic field.

8.5   Nonhomogeneous Deformation of Ferrogels

Consider a long and thin ferrogel cylinder suspended in water vertically, as shown in Figure 8.4. The magnetic field is induced by a solenoid-based electromagnet placed under the ferrogel specimen. The axis of the ferrogel cylinder (z) is parallel with the magnetic field direction and its spatial gradient. In this case, the deformation of the gel is uniaxial and can be considered as one-dimensional.

The governing equation for this situation, describing the displacement of each point of the gel along the z-axis, is the following second-order, non-linear ordinary differential eqn:[19,23-25]

$$G\left(\frac{d^2 u_z(Z)}{dZ^2} + \frac{2}{(du_z(Z)/dZ)^3}\frac{d^2 u_z(Z)}{dZ^2}\right) + M(u_z(Z))\frac{dH(u_z(Z))}{dZ} = 0 \qquad (8.36)$$

where $u_z(Z)$ represents the displacement given in the reference – undeformed – configuration, G is the shear modulus of the gel, M denotes the magnetization, and H represents the magnetic field strength.

As discussed before, the magnetization, M, of a ferrogel can be described by the Langevin function. Assuming the magnetization of individual particles in the gel to be equal to the saturation magnetization of the pure ferromagnetic material, the magnetization, M, of the ferrogel, in the presence of an applied field can be expressed by eqn (8.24), where the ξ of the Langevin function $L(\xi)$ is defined by eqn (8.25). Based on eqn (8.36) with boundary conditions $u_z(0) = 0$, $t(Z_m) = 0$, the unidirectional deformation of a ferrogel cylinder can be calculated. This deformation is shown in Figure 8.5.

The magnetic field strength along the z-axis is plotted on the left-hand side of Figure 8.5. The spatial field variation of the magnetic field in terms of z is similar to that applied in a real experiment. The gel elongates eventually, as the magnetic field intensity increases. The white lines on the gel body are indicative of the nonhomogeneous deformation of the ferrogel[19] in the nonuniform spatially varying external magnetic field. At lower field strength (*b* and *c*) the upper part of the gel elongates more than the lower part. On the other hand, at higher field strength (*d* and *e*) the lower part of the gel contracts, while the upper part elongates.

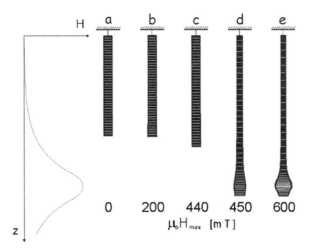

Figure 8.5 Schematic representation of the uniaxial deformation of a ferrogel cylinder calculated numerically from eqn (8.36)–(8.38). The external magnetic field distribution described by eqn (8.34)–(8.35) is also indicated in the figure. The gel on the left-hand side is undeformed ($B=0$). The next two represent the abrupt transition within a slight increase in the field intensity.
Reproduced from ref. 19 with permission from the Royal Society of Chemistry.

8.6 Concluding Remarks

The magneto–elastic properties of magnetic gels can create a wide range of deformations and motions that are controllable in the sense that varying the applied voltage generates a specific magnetic field for specific ferrogel deformation. The deformations of magnetic gels are smooth and soft, similar to mammalian muscles. The magnetically-controllable shape change of magnetic gels can be used to create an artificially designed system possessing both sensor and actuator functions internally in the gel itself. Thus, applications of magnetic gels as soft actuators for robotic and artificial muscles are well underway. The actuation and sensing of magnetic gels are noiseless and do not generate any heat, similar to Joule heating, in other electroactive artificial muscles.

Homework Problems

Homework Problem 8.1

Are shape changes in magnetic gels reversible?

Homework Problem 8.2

If the engineering stress for Neo–Hookean magnetic gels is given by:
$\sigma_{11}^{eng} = \sigma_n = G(\lambda_1 - \lambda_1^{-2})$

Show that true stress is given by: $\sigma_{11} = G(\lambda_1^2 - \lambda_1^{-1})$

Homework Problem 8.3

Describe the configuration of uniaxial magnetic gels, which will lead to a positive definite stretch.

Homework Problem 8.4

Show by a Taylor series expansion of the Langevin distribution function $L(\xi) = (\coth \, \xi - \xi^{-1})$ given in eqn (8.24) that it is approximately $L(\xi) \cong \xi/3$.

Homework Problem 8.5

Find the relationship between the magnetization, M, and magnetic field strength, H.

Homework Problem 8.6

Show that for a magnetic gel fiber with a total stretch of $l_{M,H}$

$$\lambda_{M,H}^3 - \alpha \lambda_{M,H}^2 - \beta(H_h^2 - H_m^2)\lambda_{M,H} - 1 = 0$$

where H_h and H_m are the magnetic field strengths at the bottom and the top of a rod-like ferrogel fiber, respectively (see Figure 8.5).

References

1. W. Kuhn, O. Kunzle and A. Katchalsky, Verhalten Polyvalenter Fadenmolekelionen in Losung, *Helv. Chim. Acta*, 1948, **31**, 1994–2037.
2. W. Kuhn, Reversible Dehnung und Kontraktion bei Anderung der Ionisation eines Netzwerks Polyvalenter Fadenmolekulionen, *Experientia*, 1949, **V**, 318–319.
3. A. Katchalsky, Rapid Swelling and Deswelling of Reversible Gels of Polymeric Acids by Ionization, *Experientia*, 1949, **V**, 319–320.
4. W. Kuhn, B. Hargitay, A. Katchalsky and H. Eisenberg, Reversible Dilation and Contraction by Changing the State of Ionization of High-Polymer Acid Networks, *Nature*, 1950, **165**, 514–516.
5. T. Tanaka, I. Nishio, S. T. Sun and S. Ueno-Nishio, Collapse of gels in an electric field, *Science*, 1982, **218**, 467–469.

6. Y. Osada, Conversion of the chemical into mechanical energy by synthetic polymers (chemomechanical systems), *Adv. Polym. Sci., Chap. Polym. Phys.*, 1987, **82**, 1–461.

7. D. De Rossi, K. Kawana, Y. Osada and A. Yamauchi, *Polymer Gels Fundamentals and Biomedical Applications*, Plenum Press, New York, 1991.

8. Y. Osada, Electro-Stimulated Chemomechanical System Using Polymer Gels (An Approach to Intelligent Artificial Muscle System), *Proc. Int. Conf. Intelligent Mater.*, 1992, 155–161.

9. H. J. Schneider, K. Kato and R. M. Strongin, Chemomechanical Polymers as Sensors and Actuators for Biological and Medicinal Applications, *Sensors*, 2007, **7**(8), 1578–1611.

10. M. Zrínyi and F. Horkay, Direct Observation of discrete and reversible shape transition in a magnetic field sensitive polymer gels, *J. Intelligent Mater. Syst. Struc.*, 1993, **4**(2), 190–201.

11. M. Zrínyi, L. Barsi and A. Büki, Deformation of ferrogels induced by nonuniform magnetic fields, *J. Chem. Phys.*, 1996, **104**(20), 8750–8756.

12. M. Zrínyi, L. Barsi and A. Büki, Ferrogel-A New Magneto-Controlled Elastic Medium, *Polym. Gels Networks*, 1997, **5**(5), 415–427.

13. D. Szabó, L. Barsi, A. Büki and M. Zrínyi, Studies on Nonhomogeneous Deformation in Magnetic-Field Sensitive Gels, *Models Chem.*, 1997, **134**(2–3), 155–167.

14. M. Zrínyi, L. Barsi, D. Szabó and H.-G. Kilian, Direct Observation of Abrupt Shape Transition in Hydrogels Induced by Non-Uniform Magnetic Field, *J. Chem. Phys.*, 1997, **106**(13), 5685–5692.

15. M. Zrinyi, D. Szabo and H. G. Kilian, Kinetics of the Shape Change of Magnetic Field Sensitive Polymer Gels, *Polym. Gels Networks*, 1998, **6**(6), 441–454.

16. M. Zrinyi, D. Szabo and L. Barsi, Magnetic field sensitive polymeric actuators, *J. Int. Mats. Struct.*, 1998, **9**(8), 667–671.

17. G. Torok, V. T. Lebedev, L. Cser, G. Kali and M. Zrinyi, Dynamics of PVA-gel with magnetic macro junctions, *Phys. B*, 2001, **297**(1–4), 40–44.

18. M. Zrinyi and D. Szabo, Muscular contraction mimicked by magnetic gels, *Int. J. Modern Phys. B*, 2001, **15**(6–7), 557–563.

19. M. Zrinyi, Magnetic Polymeric Gels as Intelligent Artificial Muscles, ed. M. Shahinpoor and H. J. Schneider, *Intelligent Materials*, Royal Society of Chemistry, Cambridge, UK, ch. 11, 2008.

20. R. E. Rosenzweig, *Ferrohydrodynamics*, Cambridge University Press, Cambridge/London/New York/New Rochelle/Melbourne, 1985.

21. J. I. Gittleman, B. Abeles and S. Bozowski, Superparamagnetism and relaxation effects in granular-SiO_2 and Ni-Al_2O_3 films, *Phys. Rev. B*, 1974, **9**, 3891–3897.

22. M. E. Gurtin, E. Fried and L. Anand, *The Mechanics and Thermodynamics of Continua*, Cambridge University Press, 2013.

23. M. Zrínyi, D. Szabó, G. Filipcsei and J. Fehér, Electric and Magnetic Field Sensitive Polymer Gels, in *Polymer Gels and Networks*, ed Y. Osada, A. Khokhlov and M. Decker, CHIPS, New York, 2002.

24. M. Zrinyi, D. Szab, G. Filipcsei and J. Feher, Electrical and Magnetic Field-Sensitive Smart Polymer Gels, in *Polymer Gels and Networks*, ed. Y. Osada and A. Khokhlov, Marcel Dekker, New York, 2002.

25. G. Filipcsei, I. Csetneki, A. Szilágyi and M. Zrínyi, Magnetic Field-Responsive Smart Polymer Composites, *Adv. Polym. Sci.*, 2007, **206**, 137–189.

9 Review of Electrorheological Fluids (ERFs) as Smart Material

Mohsen Shahinpoor

Mechanical Engineering Dept., University of Maine, USA
Email: shah@maine.edu

9.1 Introduction

Electrorheological fluids (ERFs) belong to a class of smart materials capable of changing from a liquid phase to a much more viscous liquid and then to an almost solid phase in the presence of an electric field. They are essentially colloidal suspensions of highly polarizable particles in a nonpolarizable solvent. The solid phase of an ERF typically has mechanical properties similar to a solid, such as a gel, and can undergo a phase change from a liquid to a thick liquid such as honey and then a solid or in reverse from a solid transform to a thick liquid and then a thin liquid in a matter of a few milliseconds. The effect is called the "**Winslow effect**", named after its discoverer Willis M. Winslow, who obtained a US patent on the effect in 1947[1] and published an article on it in 1949.[2] Note that the change is not just a change in fluid viscosity, but also the emerging solid-like properties, and hence these fluids are now known as ERFs, rather than by the older term Electro–viscous fluids (EVFs). The effect is better described as an electric field dependent shear yield stress such as what happens in a Bingham plastic (a type of viscoelastic material

Fundamentals of Smart Materials
Edited by Mohsen Shahinpoor
© The Royal Society of Chemistry 2020
Published by the Royal Society of Chemistry, www.rsc.org

such as thick honey or wax), with a shear stress yield point that is dependent on the electric field strength. ERFs once in yielding shearing mode behave like a Newtonian fluid when there is no yield shear stress and stress is directly proportional to the shear rate, γ.

There is a wide range of potential applications for ERFs, ranging from piston–cylinder type dampers to dynamic brail alphabet displays for the blind. This chapter will outline the theory by which ERFs operate, and then describe several current applications, which make use of ERFs. An ERF is typically a suspension consisting of electrically polarizable nano or micro non-conducting particles of a few microns ($\sim$4 μ) in diameter well dispersed in a non-conducting (insulating), low-viscosity medium like a colloidal solution. The particles are randomly distributed initially before an electric field is applied (Figure 9.1a). The suspended particles polarize in an applied electric field and align (Figure 9.1b) to become more solid-like, thus increasing the viscosity of the suspension when a higher strength field is applied. The viscosity changes by 100 000 times or more in response to an applied electric field of a few kilovolts per millimeter. Figure 9.1a and b depict the behavior of ERFs in an electric field.

A simple ERF can be made by mixing cornflour in a light vegetable oil or silicone oil/cross-linked polyurethane particles in silicone oil. The particles are 4 microns in diameter and contain dissolved metal ions for fast polarization and electrorheological effect. The viscosity can increase by more than five orders of magnitude under an external electric field, in the order of a few million volts per meter.

ERFs have a typical response time of a few milliseconds with reversible liquid–solid transformation under a dynamic electric field.

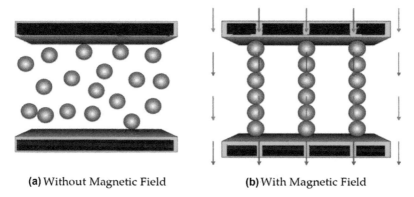

(a) Without Magnetic Field   (b) With Magnetic Field

Figure 9.1 The response of an ERF to an applied electric field.

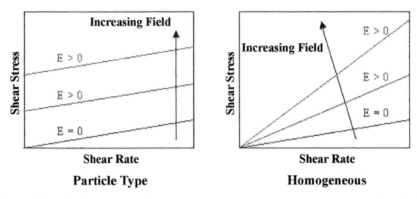

Figure 9.2 Variations in shear stress *versus* shear rates and electric fields for typical ERFs in (a) particulate form and (b) colloidal form suspensions.

After an ERF is subjected to an imposed electric field, the suspended particles will be polarized. ERFs may be in particulate or colloidal form depending on the size of particles suspended in the dielectric fluid environment.

A liquid that consists of a suspension of very fine semi-conducting particles in an electrically-insulating fluid (usually silicone oil) is an ERF suspension. When an electric current is passed through this liquid, its viscosity changes from the consistency of a liquid to that of a gel in milliseconds and reverts when the current is removed.

These ER colloidal suspensions do not settle and solidify, which could be because the makers of this substance matched the densities of the suspended particles and carrier liquid so that they would not separate, or it could be that the suspended particles are nanoparticles because they are kept suspended by 'Brownian motion,' where they randomly bump into the moving molecules of the silicone oil.

Figure 9.2a and b depict the variations in the shear stresses in ERFs as a function of electric field strength, as well as how the shear rates of deformation in particulate suspensions (a, macroparticles) and colloidal suspensions (b, nanoparticles) affect the stresses in ERFs.

For applications of these effects in microfluidics refer to ref. 3.

9.2 Giant Electrorheological Effects (GERF)

In 2003, Wen *et al.*[4] observed a new giant electrorheological fluid (GERFs) effect that enabled a much higher yield shear strength compared to normal ERFs. GERFs are made with urea-coated nanoparticles of barium titanium oxalate suspended in silicone oil.

The high yield strength is known to be due to the high dielectric constant of the particles, the nano size of the suspending colloidal particles and the urea coating of the particles before mixing. The procedure for the preparation of the suspension is given in ref. 4. In this respect, the yield stress is dependent on the electric field strength for GERFs over the full range of electric fields. If the electrodes are coated with electrically polarizable materials, a much larger increase in the Winslow ER effect can be realized, which has the effect of reducing leakage currents in the ER fluid.

9.3 Modeling of ERFs

An ERF works under a high electric field, and the particles should be polarized to such an extent that the electrostatic force between them is very strong. However, other forces act in ERFs. The structures and rheological properties of ER suspensions depend on the competition among all of the internal forces. Various dimensionless parameters have been employed to describe the relative importance of these forces. For example, the Mason number, $M_n = 6\eta\gamma/\varepsilon_0\varepsilon_{sm} (\beta E)^2$, where η_m is the viscosity of the dispersing medium, γ is the shear rate, ε_0 is the permittivity of the free space, $\beta = (\varepsilon_{sp} - \varepsilon_{sm})/(\varepsilon_{sp} + 2\,\varepsilon_{sm})$, where ε_{sp} and ε_{sm} are the static dielectric constants of the dispersing medium and dispersed particles, respectively, and E is the applied electric field. The Peclet number $\text{Peclet} = \text{Pe} = 6\eta_{sm}\pi a^2\gamma/kT$, where a is the particle radius, k_b is the Boltzmann constant, T is the temperature, indicates the ratio of the hydrodynamic to thermal forces. Dimensionless parameters that describe the relative importance of these forces are the Mason number and Peclet number such that

$$\text{Mason, } M_n = 6\eta\gamma/\varepsilon_0\varepsilon_{sm} (\beta E)^2 \tag{9.1}$$

$$\text{Peclet, Pe} = 6\eta_{sm}\pi a^2\gamma/kT \tag{9.2}$$

Four types of polarization exist, electronic, atomic, Debye, and interfacial. The dielectric constant is therefore a contribution of these, *i.e.* $\varepsilon = \varepsilon_E + \varepsilon_A + \varepsilon_D + \varepsilon_I$.

The constitutive equations for ERFs are discussed in ref. 5. These models are:

9.4 The Bingham Model

$$\tau = \tau_y + \eta_b\,\gamma \tag{9.3}$$

where τ_y is the yield shear stress, and η_b is the bulk viscosity of ERFs given by:

9.5 Krieger–Dougherty

$$\eta_b = \eta_0 \left[1 - \phi/\phi_m\right]^{-[\eta]\phi_m} \tag{9.4}$$

where ϕ is the solid volume fraction of the particles, ϕ_m is the maximum solid fraction of the particles and γ is the shear rate.

Exercise 9.1 What are the major differences between the Bingham constitutive model of ERFs and the Krieger–Dougherty model?

Answer to exercise 9.1 Bingham constitutive model is a linear model while the Krieger–Dougherty constitutive model is highly non-linear in terms of bulk viscosity and solid fractions of suspended particles.

9.6 Kinetic Chain Model

The salient features of the experimental results we have shown can be understood in terms of a kinetic model of the dynamics of volatile chains (Martin *et al.*[6]). We have presented this model elsewhere for the case of fixed induced dipolar interactions. Since the rheological properties can be easily controlled within a wide range, many scientific and technological applications may be developed. For additional references on kinetic models of ERFs refer to ref. 7–12.

9.7 Applications

ERFs can and have been used in a variety of engineering applications. These applications are highlighted in ref. 13–16. Their primary advantage is that they are capable of transforming and amplifying electrical power into mechanical effort. Because the response time of ERFs is typically in the order of milliseconds, ERFs have been successfully used as clutch systems,[16] braking systems, hydraulic valves,[13] and active damping systems.[15] Other uses, which are currently being developed, include haptic feedback[14] for prostheses such as knee or elbow prostheses. Typically, a clutch mechanism consists of a series of rotating disks which are selectively actuated to control the torque or speed of a shaft. When activated, the ERF solidifies, and the adjacent disks are forced to spin together. The ERF in the cavity between each disk may be activated independently of other cavities, and so the ERF provides a method of matching the velocity of numerous independent disks on the same shaft. The primary advantages of the ERF clutch are reduced wear and controllability. ERF braking systems work similarly, except one disk is held fixed and the activation of the ERF causes the free disk to come to a stop. Another popular application of ERFs is in hydraulic valves, which are mechanisms which regulate the flow of fluid through pipes. By using an ERF as the working fluid in a hydraulic power transmission system, the fluid flow rate may be controlled by placing electrodes throughout a pipe network and selectively activating and deactivating the electrodes to change the local viscosity, causing the working fluid to damp out any external disturbances in connection with the hydraulic valve.

9.8 Automatic Transmission and ERFs

Let us discuss the potential uses of ERFs in automatic transmissions. As will be shown, we can use ERFs in an existing automatic transmission, but using only ERF without gears is very difficult because we must somehow use different transmission ratios, which with only a fluid coupling seems impractical. The best use for ERFs is in automatic transmissions, clutches, bands, and torque converters. For an electrorheological clutch to be used in a passenger car, component dimensions similar to existing clutch plates should be chosen. Plates on the input rotor will have a radius of 105 mm.

As a good response from an ERF requires a channel width of 1 mm and a voltage of 1 kV, the radius of the housing is set at 106 mm. The width of the plate and housing section are not necessarily a factor in these equations, but are set as 2 and 4 mm, respectively. A clutch uses six plate sections in the assembly. The ERF chosen is a suspension of titanium dioxide particles in silicone oil. The relative rotational speeds of the input and output rotors at the engagement of the clutch are 4000 and 0 rpm, respectively.

As discussed before, one of the common applications of ERFs is in smart ERF clutches, as shown in Figure 9.3.

In automatic transmissions, the torque converter takes the place of the clutch found in standard shift vehicles and it is there to allow the engine to continue running when the vehicle comes to a stop. Since

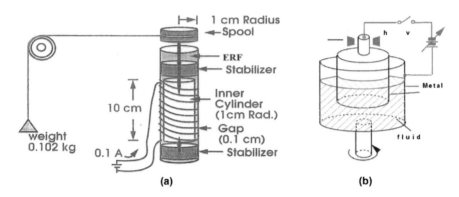

Figure 9.3   Design of (a) an ERF clutch and (b) a model of a simple ERF
clutch or brake.

the 1980s, to improve fuel economy, torque converters have been
equipped with a lockup clutch (not shown), which locks the turbine to
the pump as the vehicle speed reaches approximately 45–50 mph.
This lockup is controlled by a computer and usually does not engage
unless the transmission is in 3rd or 4th gear. If we use an ERF fluid
instead of oil in the torque converter, by applying an electric field, we
can improve the torque converter performance, which can deliver
power more rapidly and efficiently (as in with a lockup) or cut the
power stream rapidly. In terms of ERF valves and clutches, when an
electric field is applied, an ER hydraulic valve is shut, or the plates of
the clutch are locked together, when the electric field is removed, the
ER hydraulic valve is open, or the clutch plates are disengaged.

For an electrorheological clutch to be used in a passenger car,
dimensions similar to those of existing clutch plates should be
chosen. Plates on the input rotor should have a radius of 105 mm.
From a reference point on the input rotor, this translates to a relative
speed of −418 revolutions per second for the output rotor. At this
speed, the clutch transfers 3.7 watts per second from the input rotor
to the output rotor.

9.9   Conclusion

The ER effect, also known as the "**Winslow effect**", in which the viscosity of a fluid increases sharply in an electric field, was introduced.
Note that the viscosity increase is not just a change in fluid viscosity,
but also the emergence of solid-like properties, hence these fluids are
now known as ERFs, rather than by the older term, EVFs. The effect is
better described as an electric field dependent shear yield stress, such

as what happens in a Bingham plastic (a type of viscoelastic material such as thick honey or wax), with a shear stress yield point dependent on the electric field strength. The ERF once in yielding shearing mode behaves like a Newtonian fluid when there is no yield shear stress and stress is directly proportional to the shear rate, γ. There are a wide range of practical engineering applications of ERFs, many more than those listed in this summary. Interested readers should refer to the work of Martin, Litvinov, Monkman, and Winslow for additional information about ERFs, as well as the work by Carlson of the Lord Corporation, who had done much to promote both ERFs and magnetorheological fluids in recent years. The GERF effect was also discussed, which enables a much higher yield shear strength compared to that of normal ERFs. GERFs were made with urea-coated nanoparticles of barium titanium oxalate suspended in silicone oil. It turns out that the high yield strength in GERFs is due to the high dielectric constant of the particles, the nano size of the suspending colloidal particles and the urea coating of the particles before mixing. It was also noted that if the electrodes are coated with electrically polarizable materials, a much larger increase in the Winslow ER effect can be realized.

Homework Problems

Homework Problem 9.1

Design an ERF cylindrical brake system.

Homework Problem 9.2

List possible carrier fluids for ERFs.

Homework Problem 9.3

List possible particulate materials to be suspended in the ERF carriers.

References

1. W. M. Winslow, Method and means for translating electrical impulses into mechanical force, *U. S. Pat.* 2,417,850, 1947.
2. W. M. Winslow, Induced fibration of suspensions, *J. Appl. Phys.*, 1949, **20**(12), 1137–1140.
3. P. Sheng and W. Wen, Electrorheological Fluids: Mechanisms, Dynamics, and Microfluidics Applications, *Annu. Rev. Fluid Mech.*, 2012, **44**, 143–174.

4. W. Wen, X. Huang, S. Yang, K. Lu and P. Sheng, The giant electrorheological effect in suspensions of nanoparticles, *Nat. Mater.*, 2003, **2**, 727–730.
5. J. W. Goodwin, G. M. Markham and B. Vincent, Studies on Model Electrorheological Fluids, *J. Phys. Chem. B*, 1997, **101**(11), 1961–1967.
6. J. E. Martin, J. Odinek, T. C. Halsey and R. Kamien, Structure and dynamics of electrorheological fluids, *Phys. Rev.*, 1998, **57**, 756–775.
7. T. C. Halsey, Electrorheological Fluids, *Science*, 1992, **258**, 761–766.
8. D. R. Gamota and F. E. Filisko, Linear and nonlinear mechanical-properties of electrorheological materials, *Int. J. Mod. Phys. B*, 1992, **6**, 2595–2607.
9. L. C. Davis and J. M. Ginder, in *Progress in Electrorheology*, ed. K. O. Havelka and F. E. Filisko, Plenum Press, New York, 1995, pp. 107–114.
10. H. R. Ma, W. Wen, W. Y. Tam and P. Sheng, Frequency dependent electrorheological properties: origin and bounds, *Phys. Rev. Lett.*, 1996, **77**, 2499–2502.
11. W. Wen, D. Zheng and K. Tu, Experimental investigation for the time-dependent effect in electrorheological fluids under time-regulated high pulse electric field, *Rev. Sci. Instrum.*, 1998, **69**, 3573–3576.
12. W. Y. Tam, G. H. Yi and W. Wen, New electrorheological fluid: theory and experiment, *Phys. Rev. Lett.*, 1997, **78**, 2987–2990.
13. A. J. Simmonds, Electro-rheological valves in a hydraulic circuit, *Control Theor. Appl., IEEE Proc. D*, 1991, **138**(4), 400–404.
14. G. J. Monkman, An Electrorheological Tactile Display, *J. Teleoperators Virtual Environ.*, 1992, **1**(2), 219–228.
15. H. Ma, W. Wen, W. Y. Tam and P. Sheng, Dielectric electrorheological fluids: theory and experiment, *Adv. Phys.*, 2003, **52**, 343–383.
16. W. G. Litvinov, Dynamics of electrorheological clutch and a problem for a nonlinear parabolic equation with non-local boundary conditions, *IMA J. Appl. Math.*, 2008, **73**, 619–640.

10 Review of Magnetorheological Fluids as Smart Materials

Norman M. Wereley* and Young Choi

Department of Aerospace Engineering, University of Maryland, College Park, Maryland 20742, USA
*Email: wereley@umd.edu

10.1 Magnetorheological Fluids

Magnetorheological fluids (MRFs) are a family of smart materials that can transform their rheological characteristics from a fluid-like phase to a semi-solid-like phase in response to applied magnetic fields.[1–3] Typically, MRFs are composed of soft ferromagnetic particles (0.03–10 µm) dispersed in a nonmagnetic carrier fluid.

As long as the magnetizable particles exhibit low levels of magnetic coercivity, powders of many different metals and alloys can be dispersed in MRFs. Usually, the magnetic particles are pure iron, carbonyl iron or cobalt powder and the carrier fluid can be a non-magnetic organic or aqueous liquid, usually a silicone or mineral oil. In the absence of the magnetic field, the magnetic particles are randomly distributed in the fluid. However, in the presence of an applied magnetic field, the magnetic particles acquire a dipole moment aligned with the external field and form column-like chains, as shown in Figure 10.1.

Fundamentals of Smart Materials
Edited by Mohsen Shahinpoor
© The Royal Society of Chemistry 2020
Published by the Royal Society of Chemistry, www.rsc.org

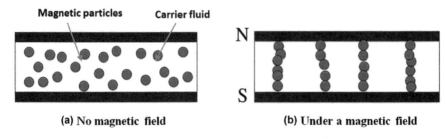

Figure 10.1 Micro-mechanism of magnetorheological fluids (MRFs).

The particle chain formation induces yield stress in the fluid that is completely reversible and also continuously and rapidly adjustable according to the intensity of the applied magnetic field. The achievable yield shear stresses of MRFs are known to be as high as 100 kPa, and the required power consumption of MRF-based devices is less than 50 W (1–2 A at 12–24 V). Also, the response time of MRFs can be as fast as 1 ms. Because of these features, MRFs have been widely applied in a variety of engineering applications, especially in adaptive vibration and shock mitigation devices such as shock absorbers, impact dampers, helicopter lag dampers, and occupant protection systems in various ground, marine, and aerial vehicles, *etc.*[4–15]

MRFs are magnetic analogs of electrorheological fluids (ERFs), which are another type of controllable yield stress fluid activated by an applied electric field.[16,17] The essential rheological characteristics of both MRFs and ERFs are similar, but there are several distinctive features. The achievable yield stress of MRFs is significantly larger than that of ERFs (*i.e.,* 5–10 kPa), and MRFs are not sensitive to impurities and contamination, and their yield stress is less dependent on the temperature, different from ERFs. However, the design and configuration of ERF-based devices can be more simple and smaller than that of MRF-based devices.

10.2   Rheological Models of MRFs

10.2.1   Bingham Plastic Model

The general rheological model for MRFs is the Bingham plastic model (see Figure 10.2)

$$\tau = \tau_y + \eta_\infty \dot{\gamma} \qquad\qquad (10.1)$$

where τ is the shear stress, τ_y is the yield stress, η_∞ is the plastic viscosity and $\dot{\gamma}$ is the shear rate. In the Bingham plastic model,[18,19]

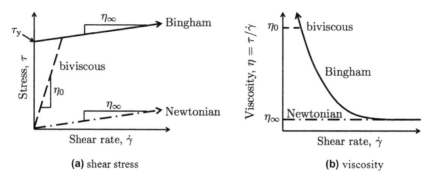

Figure 10.2 Rheological models of MRFs.

MRFs in the absence of a magnetic field show Newtonian behavior (*i.e.,* no yield stress manifests) and their shear stress is proportional to the shear rate with constant viscosity. However, when a magnetic field is applied in the Bingham model, MRFs present yield stress and the shear stress is discontinuous at zero shear rate, such that the yield stress is τ_y for positive shear rates and $-\tau_y$ for negative shear rates. This increase in the shear stress of MRFs can also be expressed by the increase in their apparent viscosity, η_a, which is defined as the instantaneous ratio of the shear stress to the shear rate as follows:

$$\eta_a = \frac{\tau_y}{\dot{\gamma}} + \eta_\infty \qquad (10.2)$$

As seen in Figure 10.2b, the apparent viscosity of MRFs in the absence of a magnetic field, is represented as a constant, η_∞. However, when the yield stress is induced by applying a magnetic field, the apparent viscosity of MRFs in the Bingham plastic flow becomes significantly larger in the lower shear rate range and is inversely exponential to the shear rate. In relatively large shear rate ranges, the apparent viscosity of the Bingham plastic flow converges to the viscosity of the Newtonian flow.

10.2.2 Biviscous Model

In the Bingham plastic model, as shown in Figure 10.2, the apparent viscosity at zero shear rate becomes infinite. Thus, the apparent viscosity of the Bingham plastic flow is discontinuous at the transient from zero shear rate to a very lower shear rate. Such discontinuous apparent viscosity of the Bingham plastic model can lead to difficulties in numerical calculations. To alleviate this discontinuity of the Bingham plastic model and regularize the pre-yield behavior through a finite viscosity, the biviscous model[20,21] has been used. In the biviscous model, the shear stress curve is constituted by two piecewise linear functions as follows:

$$\tau = \begin{cases} \tau_y + \eta_\infty \dot{\gamma}, & \text{if } \tau_y + \eta_\infty \dot{\gamma} < \eta_0 \dot{\gamma} \quad \text{(post–yield state)} \\ \eta_0 \dot{\gamma}, & \text{otherwise} \qquad\qquad \text{(pre–yield state)} \end{cases} \quad (10.3)$$

where η_0 is the pre-yield viscosity. In the biviscous model, two different yield stresses exist: one is the static yield stress, and the other is the dynamic yield stress. The static yield stress is the stress where the particle chain structures start to be broken down by an applied external shear stress, and thus, initial fluid flow begins. The static yield stress, τ_{ys}, is mathematically represented as follows:

$$\tau = \tau_{ys} \quad \text{when} \quad \tau_y + \eta_\infty \dot{\gamma} = \eta_0 \dot{\gamma} \quad (10.4)$$

Meanwhile, the dynamic yield stress, τ_{yd}, is generally determined *via* the steady-state flow curve by choosing the shear stress intercepted at zero shear rate, as follows:

$$\tau = \tau_y = \tau_{yd} \quad \text{when} \quad \dot{\gamma} = 0 \quad (10.5)$$

Thus, the dynamic yield stress implies that the minimum shear stress stops or maintains flow after initiation.

10.2.3 Shear Thickening or Shear Thinning Models

For the case where MRFs experience post-yield shear thickening or shear thinning, the Bingham plastic and biviscous models cannot capture these rheological behaviors. To solve this issue, the Herschel–Bulkley and Casson plastic models are commonly used for MRFs. The Herschel–Bulkley model[22,23] contains the power-law fluid model as follows:

$$\tau = \tau_y + K_h \dot{\gamma}^n \quad (10.6)$$

where K_h is the flow consistency index, and n is the flow behavior index. In the Herschel–Bulkley model, the effective viscosity, η_{eff}, is a function of the shear rate:

$$\eta_{\text{eff}} = K_h \dot{\gamma}^{n-1} \quad (10.7)$$

On the other hand, the Casson plastic model[24] is given by:

$$\tau^{0.5} = S_c + K_c \dot{\gamma}^{0.5} \quad (10.8)$$

where K_c is the slope of the plot of the square root of the shear stress *versus* the square root of the shear rate (*i.e.*, $\tau^{0.5}$ *versus* $\dot{\gamma}^{0.5}$), and S_c is the square root of the shear stress intercepted at zero square root shear rate. In the Casson plastic model, the yield stress is determined by:

$$\tau_y = (S_c)^2 \tag{10.9}$$

In addition, the fluid viscosity is represented as:

$$\eta = (K_c)^2 \tag{10.10}$$

10.3 Nondimensional Numbers for MRFs

10.3.1 The Bingham Number

Since the yield stress of MRFs is controllable through the application of a magnetic field, their controllability can be represented using a nondimensional Bingham number, Bi, defined by:[25,26]

$$Bi = \frac{\tau_y}{\eta\dot{\gamma}} \tag{10.11}$$

As seen in eqn (10.11), the Bingham number indicates the ratio of controllable yield stress to viscous shear stress. If using the Bingham number in eqn (10.11), the apparent viscosity in the Bingham plastic model, as shown in eqn (10.2) can be rewritten in a nondimensional form as follows:

$$\frac{\eta_a}{\eta_\infty} = 1 + Bi \tag{10.12}$$

Thus, a measurement of apparent viscosity leads directly to a measurement of the Bingham number. When the yield stress is high, or the viscosity and the shear rate are low, the Bingham number becomes large. The larger Bingham number physically implies a stronger MR effect or more controllable shear stress range. Since MRFs are used to produce controllable shear stress (or force), the Bingham number is an important nondimensional parameter for the design and analysis of MRF-based devices.

10.3.2 The Mason Number

The Mason number, Mn, is defined as the ratio of the particle shear force to particle magnetic force:[27,28]

$$Mn = \frac{16\eta_c\dot{\gamma}}{\mu_0\mu_c\beta^2 H^2} \tag{10.13}$$

where

$$\beta = \frac{\mu_p - \mu_c}{\mu_p + 2\mu_c} \tag{10.14}$$

Here, μ_0 is the permeability of the free space, $4\pi \times 10^{-7}$ [N/A^2] and H is the magnetic field, and μ_p and μ_c are the relative permeability of the particle and carrier fluid, respectively. Thus, the Mason number is used to investigate the shear behavior of particle structures in the breaking and reforming of the particle chains in MRFs. It is worth mentioning that the Mason number definition sometimes varies in the literature[29,30] because of the use of different microscopic particle interaction models and conditions. Here, the Mason number was defined using a characteristic separation distance of particle diameter in a linear particle chain model.

For MRFs, the nondimensional apparent viscosity is typically fitted to a curve of the form

$$\frac{\eta_a}{\eta_\infty} = 1 + \frac{K}{Mn} \tag{10.15}$$

where K is a characteristic constant depending on the magnetic field-induced particle structures, which can be experimentally obtained using a curve-fitting method. For a Bingham plastic model, the nondimensional apparent viscosity in eqn (10.12) is equal to eqn (10.15). This yields the relationship between the Mason number and the Bingham number for the Bingham plastic model, as follows:

$$Mn = \frac{K}{Bi} \tag{10.16}$$

In eqn (10.16), the fitted parameter K is also known as the critical Mason number, M_n^*, or the Mason number, when $Bi = 1$. MRFs with larger M_n^* (or K) values have a larger controllable yield stress ratio.

10.4 Sedimentation

For MRFs, the magnetic particles play a key role, but are also the source of sedimentation due to the large density mismatch between the magnetic particles and the carrier fluid. Such sedimentation deteriorates the designed performance of MRF-based systems and structures and can be a severe issue for certain devices that are operated after being quiescent for a long period, such as seismic dampers or impact dampers. Sedimentation is the process whereby solids separate from carrier fluids, and the solids descend to the bottom of the fluid reservoir. This process is driven by gravity due to the different specific density of the two phases. In particular, for MRFs, the difference in density between the magnetic particles (about 7.9 g mL^{-1}) and the carrier fluids (about 1 g mL^{-1}) is large. Therefore, the characterization of the sedimentation behavior of MRFs can be an essential factor in the design process of MRF-based devices.

In a column (or tube) containing an MRF with initially uniformed particle concentration, the sedimentation occurs with time and four distinct sedimentation zones (see Figure 10.3) are formed by the evolution of concentration gradients.[31–33] Since the suspended magnetic particles of MRFs tend to settle downwards, a clear liquid layer (*i.e.,* supernatant zone) is left at the top of the fluid column. The boundary or interface between the supernatant zone above and the original concentration zone below is known as the mudline. At the bottom of the fluid column, settled particles aggregate to form a highly concentrated "cake" layer (*i.e.,* the sediment zone), which is often hard to remix unless specific steps are taken to maintain stearic separation using nonmagnetic coatings. The zone that bridges the original concentration zone to the sediment zone is the variable concentration zone.

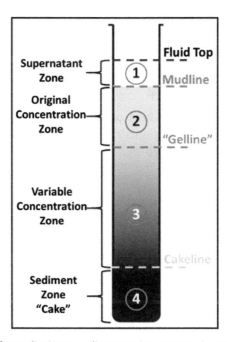

Figure 10.3 The four distinct sedimentation zones in an MRF column and their respective boundaries.

Thus, this downward progression of zones in the MRF column – from the *supernatant zone* to the *original concentration zone*, to the *variable concentration zone* to the lowest *sediment zone* – must be understood in order to develop and synthesize more stable MRFs.

Homework Problems

Homework Problem 10.1

The shear stress curve of a magnetorheological (MR) fluid and its Bingham plastic model with respect to applied magnetic fields are presented in Figure 10.4. Using the rheological properties of the MR fluid given in this problem, please answer the following questions.

(a) The yield stress of the MR fluid can be a function of magnetic field, such as $\tau_y = \alpha H^\beta$. Identify the values of α and β to be fitted with the measured yield stress by a curve-fitting method.
(b) Plot the measured yield stress data and the identified yield stress curve as a function of the applied magnetic field.
(c) Create a log–log plot for the non-dimensional Bingham number *versus* the shear rate. Use the identified yield stress values at the magnetic field inputs of 0, 50, 150, and 250 $kA\,m^{-1}$.
(d) Create a log–log plot for the apparent viscosities of the MR fluid at the magnetic field inputs of 0, 50, 150, and 250 $kA\,m^{-1}$.

Homework Problem 10.2

The mudline, z_M of an MR fluid contained in a stationary tube was measured at 2 hour intervals and the results are listed in Table 10.1. Using the sedimentation data given in this problem, please answer the following questions.

(a) Plot the measured mudline *versus* time.
(b) The measured mudline can be approximated by using two piecewise functions (the first segment at $0 \leq t \leq 12$ hours is a linear function and the second segment at $12 \leq t \leq 40$ hours is a second-degree polynomial function). Please find two piecewise functions.
(c) Obtain the mudline velocity, $\dot{z}_M$ using the identified two piecewise functions.
(d) Using the mudline velocity obtained from (c), predict the time when the sedimentation process has completed.

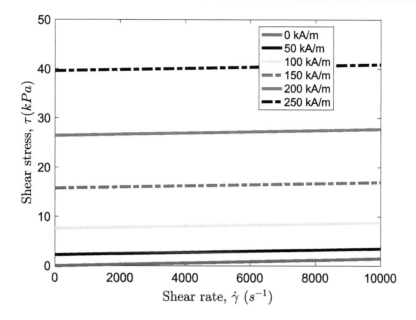

Bingham plastic model

$$\tau = 0.15 \,[\text{Pa s}]\dot{\gamma} \quad \text{at } 0 \,\text{kA m}^{-1}$$

$$\tau = 0.13 \,[\text{Pa s}]\dot{\gamma} + 2183 \,[\text{Pa}] \text{ at } 50 \,\text{kA m}^{-1}$$

$$\tau = 0.12 \,[\text{Pa s}]\dot{\gamma} + 7604 \,[\text{Pa}] \quad \text{at } 100 \,\text{kA m}^{-1}$$

$$\tau = 0.12 \,[\text{Pa s}]\dot{\gamma} + 15\,776 \,[\text{Pa}] \quad \text{at } 150 \,\text{kA m}^{-1}$$

$$\tau = 0.13 \,[\text{Pa s}]\dot{\gamma} + 26\,478 \,[\text{Pa}] \text{ at } 200 \,\text{kA m}^{-1}$$

$$\tau = 0.13 \,[\text{Pa s}]\dot{\gamma} + 39\,566 \,[\text{Pa}] \text{ at } 250 \,\text{kA m}^{-1}$$

Figure 10.4   Shear stress curve of an MR fluid and its Bingham plastic model with respect to the applied magnetic field.

Table 10.1   Measured mudline data.[a]

Time (h)	Mudline (mm)	Time (h)	Mudline (mm)	Time (h)	Mudline (mm)
0	200.0	14	187.41	28	186.05
2	198.0	16	187.21	30	185.85
4	196.0	18	187.02	32	185.66
6	194.0	20	186.82	34	185.37
8	192.0	22	186.63	36	185.26
10	190.0	24	186.44	38	185.17
12	188.0	26	186.24	40	185.08

[a]Note: The origin of the mudline, z_{M} was the bottom of the tube.

References

1. K. D. Weiss, T. G. Duclos, J. D. Carlson, M. J. Chrzan and A. J. Margida, High Strength Magneto- and Electro-Rheological Fluids, *Soc. Automot. Eng.*, 1993, 932451.
2. J. D. Carlson and K. D. Weiss, A Growing Attraction to Magnetic Fluids, *Mach. Des.*, 1994, 61–64.
3. J. D. Carlson, D. N. Catanzarite and K. A. St. Clair, Commercial Magneto-Rheological Fluid Devices, in *Proceedings of 5th International Conference on ER Fluids, MR Suspension and Associated Technology*, 1996, pp. 20–28.
4. Y. T. Choi and N. M. Wereley, Semi-Active Magnetorheological Refueling Probe Systems for Aerial Refueling Events, *Smart Mater. Struct.*, 2013, **22**, 092001.
5. M. Mao, W. Hu, Y. T. Choi, N. M. Wereley, A. L. Browne, J. Ulicny and N. L. Johnson, Nonlinear Modeling of Magnetorheological Energy Absorbers Under Impact Conditions, *Smart Mater. Struct.*, 2013, **22**, 115015.
6. G. T. Ngatu, W. Hu, N. M. Wereley and C. S. Kothera, Adaptive Snubber-Type Magnetorheological Fluid-Elastomeric Helicopter Lag Damper, *AIAA J.*, 2010, **48**, 598–610.
7. Y. T. Choi and N. M. Wereley, Biodynamic Reponses Mitigation to Shock Loads Using Magnetorheological Helicopter Crew Seat Suspensions, *J. Aircr.*, 2005, **42**, 1288–1295.
8. M. Mao, Y. Hu, Y. T. Choi and N. M. Wereley, A Magnetorheological Damper with Bifold Valves for Shock and Vibration Mitigation, *J. Intell. Mater. Syst. Struct.*, 2007, **18**, 1227–1232.
9. Y. T. Choi, R. Robinson, W. Hu, N. M. Wereley, T. S. Birchette, A. O. Bolukbasi and J. Woodhouse, Analysis and Control of a Magnetorheological Landing Gear System for a Helicopter, *J. Am. Helicopter Soc.*, 2016, **61**, 032006.
10. Y. T. Choi, C. M. Hartzell, T. Leps and N. M. Wereley, Gripping Characteristics of An Electromagnetically Activated Magnetorheological Fluid-Based Gripper, *AIP Adv.*, 2018, **8**, 056701.
11. R. Rizzo, An Innovative Multi-Gap Clutch Based on Magneto-Rheological Fluids and Electrodynamic Effects: Magnetic Design and Experimental Characterization, *Smart Mater. Struct.*, 2016, **26**, 015007.
12. W. H. Kim, J. H. Park, G.-W. Kim, C. S. Shin and S. B. Choi, Durability Investigation on Torque Control of a Magneto-Rheological Brake: Experimental Work, *Smart Mater. Struct.*, 2017, **26**, 037001.
13. Q. H. Nguyen, S. B. Choi and J. K. Woo, Optimal Design of Magnetorheological Fluid-Based Dampers for Front-Loaded Washing Machines, *Proc. Inst. Mech. Eng., Part C*, 2013, **228**, 294–306.
14. S. S. Deshmukh and G. H. McKinley, Adaptive Energy-Absorbing Materials Using Field-Responsive Fluid-Impregnated Cellular Solids, *Smart Mater. Struct.*, 2006, **16**, 106–113.
15. S. H. Choi, S. Kim, P. Kim, J. Park and S. B. Choi, A New Visual Feedback-Based Magnetorheological Haptic Master for Robot-Assisted Minimally Invasive Surgery, *Smart Mater. Struct.*, 2015, **24**, 065015.
16. M. R. Jolly, W. B. Jonathan and J. D. Carlson, Properties and Applications of Commercial Magnetorheological Fluids, *J. Intell. Mater. Syst. Struct.*, 1999, **10**, 5–13.
17. Y. T. Choi and N. M. Wereley, Comparative Analysis of the Time Response of Electrorheological and Magnetorheological Dampers Using Nondimensional Parameters, *J. Intell. Mater. Syst. Struct.*, 2002, **13**, 443–451.
18. R. W. Phillips, Engineering Applications of Fluids with a Variable Yield Stress, Ph.D. Dissertation, University of California, Berkeley, 1969.
19. S. G. Sherman, Magnetorheological Fluid Dynamics for High Speed Energy Absorbers, Ph.D. Dissertation, University of Maryland, College Park, 2016.

20. R. A. Synder, G. M. Kamath and N. M. Wereley, Characterization and Analysis of Magnetorheological Damper Behavior Under Sinusoidal Loading, *AIAA J.*, 2001, **39**, 1240–1253.

21. N. M. Wereley, J. U. Cho, Y. T. Choi and S. B. Choi, Magnetorheological Dampers In Shear Mode, *Smart Mater. Struct.*, 2008, **17**, 015022.

22. A. Chaudhuria, N. M. Wereley, S. Kothab, R. Radhakrishnanb and N. M. Sudarshanb, Viscometric Characterization of Cobalt Nanoparticle-Based Magnetorheological Fluids Using Genetic Algorithms, *J. Magn. Magn. Mater.*, 2005, **293**, 206–214.

23. N. M. Wereley, Nondimensional Herschel-Bulkley Analysis of Magnetorheological and Electrorheological Dampers, *J. Intell. Mater. Syst. Struct.*, 2008, **19**, 257–268.

24. M. A. Rao, *Rheology of Fluid, Semisolid, and Solid Foods: Principles and Applications, Food Engineering Series*, Springer US, New York, 2014.

25. N. M. Wereley and L. Pang, Nondimensional Analysis of Semi- Active Electrorheological and Magnetorheological Dampers Using Approximate Parallel Plate Models, *Smart Mater. Struct.*, 1998, **7**, 732–743.

26. S.-R. Hong, N. M. Wereley, Y. T. Choi and S. B. Choi, Analytical and Experimental Validation of a Nondimensional Bingham Model for Mixed Mode Magnetorheological Dampers, *J. Sound Vib.*, 2008, **312**, 399–417.

27. A. C. Becnel, S. G. Sherman, W. Hu and N. M. Wereley, Nondimensional Scaling of Magnetorheological Rotary Shear Mode Devices Using the Mason Number, *J. Magn. Magn. Mater.*, 2015, **380**, 90–97.

28. S. G. Sherman, A. C. Becnel and N. M. Wereley, Relating Mason Number to Bingham Number in Magnetorheological Fluids, *J. Magn. Magn. Mater.*, 2015, **380**, 98–104.

29. D. Klingenberg, J. Ulicny and M. Golden, Mason Numbers for Magnetorheology, *J. Rheol.*, 2007, **51**, 883–893.

30. M. T. López-López, J. D. G. Durán, Á. V. Delgado and F. González-Caballero, Scaling Between Viscosity and Hydrodynamic/Magnetic Forces in Magnetic Fluids, *Croat. Chem. Acta*, 2007, **80**, 446–451.

31. L. Xie, Y. T. Choi, C.-R. Liao and N. M. Wereley, Characterization of Stratification for an Opaque Highly Stable Magnetorheological Fluid Using Vertical Axis Inductance Monitoring System, *J. Appl. Phys.*, 2015, **117**, 17C754.

32. Y. T. Choi, L. Xie and N. M. Wereley, Testing and Analysis of Magnetorheological Fluid Sedimentation In a Column Using a Vertical Axis Inductance Monitoring System, *Smart Mater. Struct.*, 2016, **25**, 04LT01.

33. J. M. Chambers and N. M. Wereley, Vertical Axis Inductance Monitoring System to Measure Stratification In a Column of Magnetorheological Fluid, *IEEE Trans. Magn.*, 2017, **53**, 4600205.

11 Review of Dielectric Elastomers (DEs) as Smart Materials

Mohsen Shahinpoor

Mechanical Engineering Dept., University of Maine, USA
Email: shah@maine.edu

11.1 Introduction

If rubbery elastomers such as a silicone rubber sheet are sandwiched between two compliant electrodes, then any imposed electric field induces electrostatic forces (attraction) between the electrodes. Thus, the rubber sheet in between them can be compressed by the electrostatic forces, which then cause the rubbery sheet to expand sideways due to the Poisson's ratio effect and thus actuation results. In 1880, Röntgen[1] demonstrated this actuation using two glasses as dielectrics, and once the opposing surfaces of these glasses were charged, small thickness changes were observed. Later electrostatically-induced pressures acting to compress dielectrics became known as the "Maxwell stress." It was, however, Pelrine, Kornbluh, and Joseph in 1998[2] who introduced dielectric elastomer technology with compliant electrodes. They concluded that by deliberately choosing polymers with relatively low moduli of elasticity, the field-induced strain response due to Maxwell stress could be large. Polymer actuators designed to exploit Maxwell stress in this manner became known as dielectric elastomers, or Des, following several

Fundamentals of Smart Materials
Edited by Mohsen Shahinpoor
© The Royal Society of Chemistry 2020
Published by the Royal Society of Chemistry, www.rsc.org

pioneering papers by Pelrine, Kornbluh, and co-workers,[3-6] and Wax and Sands.[7]

In this chapter, the basics of dielectric elastomers are reviewed from materials and operational perspectives, and some basic designs of DE actuators, generators and sensors are discussed for a variety of applications.

11.2 Fundamentals of Dielectric Elastomer Actuation

Dielectric elastomer actuators (DEAs) are smart materials made with soft dielectric materials such as silicone rubber, which produces large strains under pressure. They have been classified as electroactive polymers (EAPs). Based on their simple working principles, dielectric elastomer actuators (DEAs) directly transform electrical energy into mechanical work. DEAs are lightweight and have a high elastic energy density. Figure 11.1 displays the basic mechanism of actuation in dielectric elastomer actuators (DEAs).

The compliant electrodes are mounted on the surfaces of the sandwiched dielectric elastomer by various deposition techniques, such as physical vapor, chemical vapor, or ion implantation.

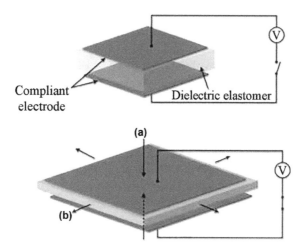

Figure 11.1 Actuation principle of DEAs: (a) dielectric elastomer sandwiched between two compliant electrodes capable of stretching out with the elastomer, (b) dielectric elastomer squeezed between the compliant electrodes due to the imposed voltage and electrostatic attraction between the compliant electrodes.

When a voltage, V, is applied, the electrostatic pressure, pel, arising from the Coulombic forces acting between the electrodes normally squeeze and compress the sandwiched elastomer to deform it outward and make it thinner, as shown in Figure 11.1.

The equivalent electrostatic pressure, pel, or the Maxwell stress is given by the following equation:

$$p_{el} = \varepsilon_r \varepsilon_0 E^2 \qquad (11.1)$$

where ε_0 is the vacuum permittivity $(8.85 \times 10^{-12} F/m)$, ε_r is the relative permittivity or the dielectric constant of the polymer, $E = (V/t)$ is the electric field with the imposed voltage across the thickness, t, of the DEA film. For the elastomer, often silicones and acrylic elastomers are used. These elastomers have a Young's modulus of about 0.5–1 MPa and a relative dielectric constant of 3. A high dielectric constant, high electric breakdown field and a low Young's modulus all contribute to an increase in strain.

It has been a challenge to find elastomers which have a favorable combination of these parameters, and often the improvement of one property will result in a trade-off of the other two properties. Note that the stress in eqn (11.1) is twice the classical Maxwell stress present in capacitors with rigid electrodes because the electrodes here are compliant.[2–6] In particular, the acrylic elastomer VHB 4910, commercially available from the company 3M, has shown the largest activation strain, a high elastic energy density, and high electrical breakdown strength. The desirability of pre-stretching the elastomer before the application of a high voltage, as well as the occurrence of dielectric breakdown during actuation, have been major issues in connection with DEA development. Pre-stretching is desirable to reduce the initial high voltage for electrostatic attraction. Of course, the high voltage requirement for DEAs has also been an issue in connection with handling high voltages in the range of a few thousands of volts. If the dielectric elastomer deformation is not infinitesimal, which is normally the case, the Maxwell stress given in eqn (11.1) is nonlinearly related to the strain, s, or $p_{el} = f(s)$, where $f(s)$ is a nonlinear function of s, and can be assumed to have either Neo–Hookean or Mooney–Rivlin type constitutive equations, as discussed by Treloar[8] and Hoss and Marczak.[9]

A more detailed account of how to combine the theory of hyperelasticity with Maxwell's stress is presented later in this chapter. According to Pelrine, Kornbluh, Pei and Joseph,[3] pre-stretching dielectric elastomers before actuation increases the maximum possible strain under reasonable electric fields of a few MV/m such that with small imposed biaxial pre-strains the maximum area strain achieved was just 40%, while with high pre-strains, the same film achieved an area strain of 158%.[3] This increase in maximum strain is mostly due to the electrical breakdown strength of the polymer, but the mechanism behind this increase is still unclear. Pre-stress and pre-strain will be used interchangeably in this chapter.

11.3 The Challenge of Mounting Compliant Electrodes on DEAs

Compliant electrodes are extremely critical to the operation of dielectric elastomers. The elastomers can undergo large elongations, and so must the electrodes. Thus, new approaches to making such compliant electrodes have been developed. Percolating networks of conducting particles in a matrix have been the most commonly used approach, according to Pelrine, Kornbluh, and Joseph.[2] The low percolation threshold is desirable in this case to enhance and increase the matrix conductivity by using more conductive particles in the compliant matrix. Carbon grease electrodes can also be directly produced from silicone oil and carbon black. The carbon grease is also helpful for quick demonstrations of new actuator designs. However, the grease easily rubs off, and grease-based electrodes have been shown to lower the performance of DEAs, according to Carpi, Chiarelli, Mazzoldi and de Rossi.[10] More solid elastomeric electrodes may also be fabricated by mixing of a compliant matrix substrate with percolating conducting fillers.[11] Initial attempts have been made to employ MEMS technologies in which UV lithography may be used to define thin zig-zag electrodes, which retain conductivity even if the underlying material undergoes large linear deformation. Similarly, corrugated electrodes, which are prepared by direct deposition of a metal film on top of a silicon film with a corrugated surface, may be used as compliant electrodes.

As recently reported by Keplinger, Kaltenbrunner, Arnold, and Bauer[12] electrical actuators made from films of dielectric elastomers coated on both sides with stretchable electrodes have potential applications as soft biomimetic actuators but suffer from pull-in electromechanical instability and dielectric breakdown, limiting operational voltages and attainable deformations. In 1880, Röntgen[1] proposed and studied electrode-free actuators driven by sprayed-on electrical charges. In this free-electrode configuration, the DEAs withstand much higher voltages and deformations and allow for electrically clamped (charge-controlled) thermodynamic states preventing electromechanical instabilities. Another great advantage of the electrode-free actuation of DEAs allows the direct optical monitoring of the actuated elastomer, as well as the design of new 3D actuator configurations and adaptive optical elements.[12]

11.4 Constitutive Equations for Dielectric Elastomer Actuators

According to Martins, Natal Jorge, and Ferreira,[13] the stress–strain or stress–stretch (l) relationship of a hyperelastic material can be described by several models such as Neo–Hookean[14] and Mooney-Rivlin[15] constitutive equations. Generally speaking, the stresses are derivable from the strain energy function W, which is a function of the strain tensor, s_{ij}, or its invariants, I_1, I_2, and I_3, such that:

$$W = W (I_1, I_2, I_3) \tag{11.2}$$

In this work, the strain energy expressions are particularized for incompressible materials only ($I_3 = 1$), so that the strain energy function takes the form $W = W (I_1, I_2)$. In what follows, it is assumed that the material is incompressible and environmental parameters such as temperature, dielectrics, incompressibility, and isotropic parameters are constant during the deformation modeling. With these assumptions, the first and second invariants of the strain tensor can be calculated as:

$$I_1 = \lambda_1^2 + \lambda_2^2 + \lambda_3^2 \tag{11.3}$$

$$I_2 = 1/\lambda_1^2 + 1/\lambda_2^2 + 1/\lambda_3^2 \tag{11.4}$$

$$I_3 = \lambda_1 \lambda_2 \lambda_3 = 1 \text{ (incompressibility)} \tag{11.5}$$

where, λ_1, λ_2 and λ_3 are the stretch ratio of the DEA material (Figure 11.2) such that in the longitudinal, width and thickness directions, respectively, the strain energy functions can be presented as:

$$W = C_1(I_1 - 3): \text{ (for \textbf{Neo–Hookean} DEAs,}^{13} \text{ up to 100\% strain),} \tag{11.6}$$

$$W = C_1(I_1 - 3) + C_2(I_2 - 3): \text{ (\textbf{Mooney–Rivlin} DEAs,}^{15} \text{ strain} > 100\%),} \tag{11.7}$$

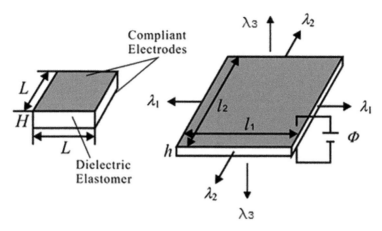

Figure 11.2 Dimensions L, H, l_1, l_2, $h = l_3$ and stretches λ_1, λ_2 and λ_3 in a DE actuator.

where C_1 and C_2 are materials parameters that can be determined using fitting data of the experimental results.

When strains are 100% or less, then the Neo–Hookean material model presents a very good fit for the mechanical behavior. Otherwise, for larger strains, the Mooney–Rivlin model is recommended.[2–4,15,16] Typical values of $C_1 = G/2$, which is also related to the shear modulus of hyperelasticity, or G, are less than or equal to 0.5 MPa, making the typical modulus of the elasticity of Neo–Hookean hyperelastic materials $E = 3G$ due to incompressibility. Otherwise,

$$G = (E/2(1 + \nu))\qquad(11.8)$$

where ν is the Poisson's ratio, which turns out to be equal to (1/2) for incompressible materials. A typical value of C_2 for Mooney–Rivlin materials is around 0.4 MPa.[13–15]

Making an assumption for dielectric elastomers, such as an in-compressible, isotropic, and homogeneous hyper-elastic Neo–Hookean model, the true stress formula can be presented as:[15]

$$\sigma_{ij} = \lambda_i \frac{\partial W}{\partial \lambda_j} - p\delta_{ij}\qquad(11.9)$$

so that

$$\sigma_{11} = \lambda_1 \frac{\partial W}{\partial \lambda_1} - p, \; \sigma_{22} = \lambda_2 \frac{\partial W}{\partial \lambda_2} - p, \; \sigma_{33} = \lambda_3 \frac{\partial W}{\partial \lambda_3} - p\qquad(11.10)$$

Note that W is a function of I_1 and I_2 and thus by chain rule of differentiation, one can find the following expressions for stresses in DEAs. Note that in this case, due to incompressibility, $\lambda_1\lambda_2\lambda_3 = 1$, and thus, for a uniform plate of DEA (Figure 11.2) $\lambda_2 = \lambda_1 = \lambda_3^{(-1/2)} = \lambda^{(-1/2)}$ and thus:

$$W = C_1(\lambda_1^2 + \lambda_2^2 + \lambda_3^2 - 3) + C_2(\lambda_1^{-2} + \lambda_2^{-2} + \lambda_3^{-2} - 3) \tag{11.11}$$

or simply:

$$W = C_1(\lambda_3^2 + 2\lambda_3^{-1} - 3) + C_2(\lambda_3^{-2} + 2\lambda_3 - 3) \tag{11.12}$$

Thus, it can be deduced from eqn (11.10)–(11.12) that:

$$\sigma_{33} - \sigma_{22} = \lambda_3 \frac{\partial W}{\partial \lambda_3} - \lambda_2 \frac{\partial W}{\partial \lambda_2}, \quad \sigma_{22} - \sigma_{11} = \lambda_2 \frac{\partial W}{\partial \lambda_2} - \lambda_1 \frac{\partial W}{\partial \lambda_1} \tag{11.13}$$

Since in this case, $\sigma_{22} = \sigma_{11} = 0$, the expression for the engineering Maxwell stress, σ_{33}, is derived such that:

$$\sigma_{33} = \lambda_3 \frac{\partial W}{\partial \lambda_3} - \lambda_2 \frac{\partial W}{\partial \lambda_2} = 2C_1(\lambda_3^2 - \lambda_3^{-1}) - 2C_2(\lambda_3 - \lambda_3^{-2}) \tag{11.14}$$

For Mooney–Rivlin-type DEAs.[15] Note that σ_{33} is the true stress or axial force applied to the DEA divided by the deformed cross section of the plate, which in this case is related to the initial unreformed cross section by the product of λ_1 and λ_2, which due to incompressibility (eqn (11.5)), is equal to $\lambda_1\lambda_2 = \lambda_3^{-1} = \lambda^{-1}$. Thus, the engineering stress $\sigma_{33}^{eng} = \sigma_n = \lambda_3^{-1}\sigma_{33} = \lambda^{-1}\sigma_{33}$ in the dielectric elastomer actuator in the z-direction can be expressed as:

$$\sigma_n = 2(C_1 - C_2\lambda^{-1})(\lambda - \lambda^{-2}) \tag{11.15}$$

where the nominal stress, σ_n, is defined as the ratio of the equilibrium elastic force to the undeformed cross-sectional area of the sample and λ is the stretch in the z-direction.

Exercise 11.1 Derive eqn (11.15) from the generalized constitutive equations for rubber elasticity.

The solution to Exercise 11.1: From eqn (11.14), we note that:

$$\sigma_{33} = 2C_1(\lambda_3^2 - \lambda_3^{-1}) - 2C_2(\lambda_3 - \lambda_3^{-2}) = 2C_1(\lambda^2 - \lambda^{-1}) - 2C_2(\lambda - \lambda^{-2}) \tag{11.16}$$

The expression (11.16) is the actual stress $= F/A$, where F is the electrostatic attraction force, and A is the deformed cross section under the compliant electrode. Note that this deformed cross section is related to the undeformed cross section of the dielectric elastomer actuator or A_0 such that $A = \lambda_1\lambda_2 A_0 = (1/\lambda_3) A_0$, due to incompressibility. Finally one has $A = (1/\lambda) A_0$. Thus,

$$\sigma_n = \lambda_3^{-1}\sigma_{33} = 2\lambda_3^{-1}C_1(\lambda_3^2 - \lambda_3^{-1}) - 2C_2\lambda_3^{-1}(\lambda_3 - \lambda_3^{-2}) = 2(C_1 - C_2\lambda^{-1})(\lambda - \lambda^{-2})$$

$$(11.17)$$

The simplicity of the Neo–Hookean model permits easy evaluation, but unfortunately, it only holds well for strains to at most 100%.[6] Simply stated, the elastic engineering stress tensor is proportional to the stretch, λ_3, such that

$$\sigma_n = G(\lambda_3 - \lambda_3^{-2}) \qquad (11.18)$$

where λ_3 is the deformation ratio in the thickness direction and G is the shear modulus, which can be measured directly in stress–strain experiments.

The above analysis applies to the situation when the actuator carries no load. The analysis can be expanded to the loaded situation, reflecting the situation for the simplest planar actuator, capable of outputting useful actuation. The simplest case is when the width of the actuator is fixed, $x_2 = x_2'$ (the 2-direction is along the width). This constraint makes the length and thickness deformation ratios inversely proportional, $\lambda_1 = \lambda_3^{-1}$. The load is taken to be due to a constant mass, m, attached to the bottom of the actuator. Then, the tension on the face of the elastomer in the 1-direction is $T_{11} = mg/x_2x_3 = mg/x_2'x_3'\lambda_3$, where x_2x_3 is the end face area in the 1-direction. Combined with the above, the result is

$$\lambda_1^2 \frac{\epsilon_r\epsilon_0 V^2}{x_3'^2} + \lambda_1 \frac{mg}{x_2'x_3'} = G(\lambda_1 - \lambda_1^{-2}) \qquad (11.19)$$

which can be solved numerically for each set of mass and voltage chosen. The actuation strain is defined as the length of the actuator at a given voltage and mass with respect to the un-actuated length at the same mass,

$$\varepsilon^{act} = \frac{x_1(m, V) - x_1(m, 0)}{x_1(m, 0)} = \frac{\lambda_1(m, V) - \lambda_1(m, 0)}{\lambda_1(m, 0)} \qquad (11.20)$$

and is a practical way to present the actuation performance of any linear actuator.

The deformation ratios and actuation strains calculated for a typical actuator ($G = 100$ kP$_a$, $\epsilon = 2.5$, $x_2' = 100$ mm, $x_3' = 100$ μm) are shown in Figure 11.3. Although the strains that refer to the initial

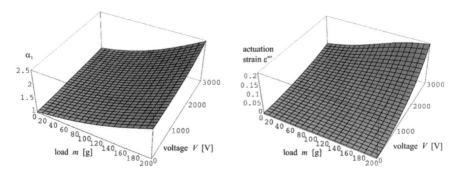

Figure 11.3   Left: deformation ratio of a flat film actuator, based upon the
Mooney–Rivlin model. Right: actuation strain for the same
model.
Reproduced from ref. 6 with permission from the Royal Society
of Chemistry.

unloaded, zero voltage state are high, the achieved actuation strains
are not. In this rather simple model (which underestimates the elastic
stress because it neglects strain hardening), it is seen that the actuation strain at a given voltage can be optimized by changing the load.

A model for the description of the actuation of a tube was presented
by Carpi and de Rossi[17] and Carpi, Migliore, Serra and D. Rossi.[18] The
measured actuation strains were in this case quite small and per-
mitted the use of the Neo–Hookean constitutive equations. It was
shown how the model, again with no free parameters, accounted well
for the experimental actuation curves obtained on a silicone elastomer, a medical tube with grease electrodes.

To conclude this section, models which are meant to describe the
large strain and actuation capabilities of dielectric elastomer actuators must include hyper-elastic formulations of the constitutive
equations. Most of our discussion so far has been based on quasi-
static models of performance. The often large components of the
visco-elastic loss of some polymers (particularly acrylics) require that
a theoretical description must be based on hyper-elastic constitutive
equations with a visco-elastic part, to properly account for the time-
dependent behavior of actuation. Finally, for a complete analysis of a
more complex actuator structure, a finite element approach is often
unavoidable.

Exercise 11.2: Referring to Figure 11.4, design a cylindrical dielectric elastomer actuator.

Determining the dynamics of the actuator design requires the
governing equations. The simplistic model will be based on the parallel plate capacitor model.

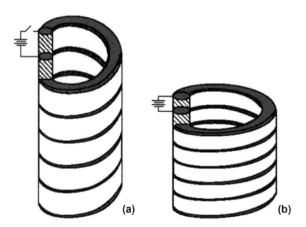

Figure 11.4 Helical dielectric elastomer actuator, (a) undeformed and (b) deformed.

Table 11.1 Typical properties of DEA materials (Source: SoftRobotics Toolkit, http://softroboticstoolkit.com).

Elastomer Material	Film Thickness (micron, μ)	Strain at Break	Young's Modulus at 50% (MPa)	Creep [%]
Polyurethane TRU LPT 4210	50	421	3.36	24
Polyurethane Bayfol EA102	50	300	1.44	2.9
Silicone	45	422	0.25	16
Acrylate 3M VHB 4905	498	879	0.04	70

Table 11.1 depicts some materials properties for DEA soft polymers, including the properties of Acrylate 3M VHB 4905.

Geometry of actuator, thickness $= 0.8$ mm, width $= 8$ mm, length $= 100$ mm, and $Y_E = 0.04$ MPa. Assuming the material is isochoric and acts in a Neo–Hookean manner, this will accurately model the actuator for the lower end of the stress. This actuator will compress, as seen in Figure 11.4. Note that:

$E = (V/l)$ and the Maxwell stress $\sigma_\nu = \varepsilon_0 \, \varepsilon_\rho E^2$, where σ_n is the Maxwell pressure, ε_0 is the vacuum permittivity, ε_r is the dielectric constant of the polymer, V is the voltage, and Y_E is the modulus of elasticity. A Neo–Hookean model can be determined in the following way:

$$\sigma_m = (Y_E/3)(\lambda - \lambda^{-2}) \tag{11.21}$$

where λ is the stretch or $\lambda = (L/L_0)$, where L and L_0 are the current and initial thicknesses of the DEA that determine the change in length from 0–4000 V.

MATLAB Solution to Exercise 11.2: Here is a MATLAB code to solve the problem and plot the results:

```
t= 0.0008;%m
w =0.008;%m
L_0=0.100;%mm
Y_E=0.04E6;% pa
eps_0=8.85418782e-12;
eps=7;
syms L
i=1;
for V=1:20:4000
    Voltage(i)=V;
    if V==1;
    E(i)=V/t;
    else
        E(i)=V/(t*outs(1,i-1)/L_0);
    end
    P(i)=eps_0*eps*E(i)^2;
    outs(:,i)=double(solve(-P(i)==Y_E/3*
        (L/L_0-(L/L_0)^-2)));
    DelLength(i)=outs(1,i)-L_0;
    i=i+1;
end
figure(5)
plot(Voltage,DelLength*-1000)
title('Figure 5: Shrinking of Actuator')
xlabel('Voltage(V)')
ylabel('Length (mm)')
figure(6)
plot(Voltage,E/1000)
title('Figure 6: Electric Field vs Voltage')
xlabel('Voltage (V)')
ylabel('Electric Field (kV/m)')
```

Figure 11.5 displays the variations in the length or dimensions of the DE with respect to the applied voltage.

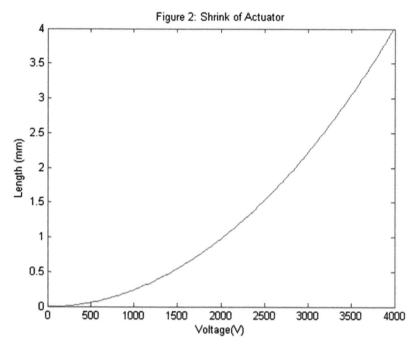

Figure 2: Shrink of Actuator

Figure 11.5 Variations in the contraction length *versus* the applied voltage.

11.5 Actuator Design: Geometry and Structure

The basic functional element of a dielectric elastomer actuator is the elastomeric film sandwiched between compliant electrodes, as shown in Figure 11.1. This basic element must have both mechanical and electrical connections. Mechanical connections are particularly critical since the polymer materials are relatively soft and poor design could result in little energy being coupled to the load. The dielectric elastomer material is essentially incompressible. In the functional element, the electrodes uniformly impart effective compressive stress in the z-direction, which tends to decrease the thickness of the polymer film. The incompressible polymer film expands in the area so that the total volume of polymer between the electrodes is conserved. Depending on the loading and constraints, the area expansion can occur equally in both planar directions or only in a single direction. In many applications, motion is desired only in a single planar direction. Undesirable deformation can be restrained by adding anisotropy into the material, or by changing the geometry, such as making the active area much wider than it is long.

Note that it is not always necessary to couple two directions of planar expansion into a single direction of the output. For example, in diaphragm actuators that operate across a pressure gradient such as might be used for a pump, the two directions of planar deformation are implicitly coupled to the output. In this case, the planar deformations displace the diaphragm and thereby impart momentum to the fluid being pumped. In some instances, the output of the actuator is produced by only a change in the geometry of a material or coating, with little or no force applied. In some cases, it is desirable to expand the dielectric elastomer uniformly in both planar directions, such as in a solid-state optical aperture. Note that the pressure that is produced in a dielectric elastomer, as given by eqn (11.2), does not depend on the elasticity of the polymer. Thus, while the dielectric elastomer materials themselves are relatively soft, the pressure or force need not be small.

If the elastomer is properly coupled to the load, then the pressure or forces that can be produced are determined by the strength of the electric field that can be produced in the material. In the following section, we highlight different applications of dielectric elastomer actuators. The various applications selected also highlight a variety of different actuator configurations. How these actuator configurations deal with the loading issues are discussed.

11.6   Artificial Muscles for Biomimetic Robots

A variety of actuator configurations have been explored for use as artificial muscles in robotic applications. Rolled actuators have high force and stroke in a relatively compact package. Because, like a muscle, this package is cylindrical, rolled actuators are an obvious choice for artificial muscles. One of the more successful types of rolled actuators is the "spring roll" (see Figure 11.6).

This actuator is formed from an acrylic film that is stretched in tension and rolled around an internal spring. Rolled actuators have been made in a variety of lengths and diameters with maximum stokes of up to 2 cm and forces of up to 33 N. Figure 11.4 shows a rolled dielectric elastomer actuator that weighs approximately 15 g and can produce peak forces of more than 5 N with maximum displacements of 5 mm (a strain of about 25%).

The maximum performance of dielectric elastomer devices is expected to increase as designs are improved since these devices are still well below the measured peak performance of the polymer materials.

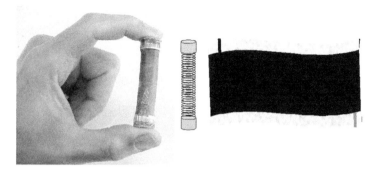

Figure 11.6   Spring roll linear actuator, where the structure of the actuator is shown on the right.
Reproduced from ref. 6 with permission from the Royal Society of Chemistry.

The main disadvantage of the rolled actuator is that it is somewhat more difficult to fabricate since it is no longer a flat structure. Additionally, the energy coupling of a long roll is not as good as that of the bow-tie or trench type of actuator, since only one direction of deformation is coupled to the load. The energy coupling of short rolls is however, better.

11.7   DE Sensors

We have already noted that dielectric elastomer actuators are also intrinsically positioned or strain sensors. It has been shown, for example, that measurement of the capacitance can indicate the displacement of a rolled actuator.[6] Dielectric elastomer sensors have certain advantages in their own right, even if the actuation function is not included. The large strain capabilities and environmental tolerance of dielectric elastomer materials allow for sensors that are very simple and robust. In sensor mode, it is not often important to maximize the energy density of the sensor materials, since relatively small amounts of energy are converted. Thus, the selection of dielectric materials can be based on criteria such as biocompatibility, maximum strain, environmental survivability, and even cost. As sensors, dielectric elastomers can be used in all of the same configurations as actuators, as well as others. Figure 11.7 shows examples of several dielectric elastomer sensors, which can be thin tubes, flat strips, arrays of diaphragms, or large-area sheets.

In many applications, these sensors can replace bulkier and more costly devices such as potentiometers and encoders. In fiber or ribbon

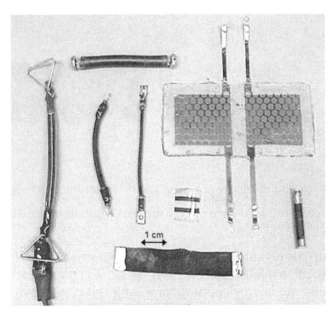

Figure 11.7 Representative dielectric elastomer sensors.
Reproduced from ref. 6 with permission from the Royal Society
of Chemistry.

form, a dielectric elastomer sensor can be woven into textiles and
provide position feedback from human motion. Such sensors are
capable of measuring respiration, replacing bulkier pressure sensing
chest bands. The use of such strain sensors has also been explored for
measuring the expansion of arteries in animal models. These simple
one-piece sensors can be made very small and thin. They can replace
saline- or mercury-filled tubes that are commonly used by physi-
ologists to measure the expansion of body parts. The softness and
compliance of dielectric elastomers are ideal for interaction with the
human body. Dielectric elastomer sensors can also be laminated to
structures or skins to provide position information for multi-
functional smart materials. Sensors such as the diaphragm array can
be used to measure force or pressure as well as motion. Enhanced
thickness mode pads can act as sensory skins.

11.8 The Future: Materials Development for New Elastomers

The characteristics of elastomer dielectrics are critical to the performance of this actuator technology. The majority of research has

thus far focused on the use of existing commercially available polymers. However, the development of new polymers or polymer formulations specifically optimized for use as dielectric elastomers are beginning to emerge and could allow for dramatic improvements in the technology in years to come. It is therefore fitting that we conclude this chapter with a look to the future of materials development.

As already mentioned, the dielectric elastomer serves several purposes for the dielectric elastomer actuator. Both its electrical and mechanical properties are important; hence, there are several approaches to improving actuator performance by varying the properties of the dielectric elastomer. It must be emphasized, however, that it is not likely that one particular dielectric elastomer material will be optimal for all applications. Rather, the large number of possible applications demands that a variety of actuation strains, stresses, speed responses, viscoelastic losses, and their combinations are possible. Typically, low mechanical loss is desired, but for some applications, inherent passive damping or shock absorption is necessary, which can then be facilitated by manipulation of the visco-elastic loss spectrum. Currently, researchers are attempting to expand the library of materials available for applications, most often basing their approach on the two hitherto most successful elastomers, polydimethylsiloxane (silicone) rubber and polyacrylate rubber. For additional discussions on how to improve the elastic, dielectric, and breakdown characteristics, refer to the work by Kofod and Kornbluh in ref. 6

11.9 Conclusions

The past 15 years have seen the dielectric elastomer actuator concept evolve from an idea to a commercially available product, carried through by major efforts in materials development and discovery, actuator design, modeling and theory, and applications studies. We can expect even greater growth in the field in the next few years.

The unique capabilities of dielectric elastomers should allow for their use in a wide variety of applications, not just as replacements for existing actuators, but as enablers of new devices and systems. Realizing this vision will require addressing lifetime issues, manufacturability, and systems issues, such as electrical requirements. Successful innovation does not require any fundamental improvement in performance. Nonetheless, great advances are still possible. Many groups are developing new materials with drastically improved mechanical and dielectric properties. New actuator designs are constantly being developed. Perhaps most importantly, new applications that exploit the unique capabilities of dielectric elastomers are constantly being explored. With these developments, the goal of providing compact, light-weight, and powerful artificial muscles is within reach.

Homework Problems

Homework Problem 11.1

Derive eqn (11.13) from the generalized constitutive equations for rubber elasticity.

Homework Problem 11.2

Design of a helical dielectric elastomer actuator.

Geometry of the actuator, thickness $= 0.8$ mm, Width $= 8$ mm, Length $= 100$ mm, $Y_E = 0.04$ MPa. Assuming the material is isochoric and acts in a Neo–Hookean manner, the actuator for the lower end of the stress will be accurately modeled. This actuator will compress, as seen in Figure 11.4. Note that $E = (V/l)$ and the Maxwell stress $\sigma_\nu = \varepsilon_0 \varepsilon_\rho E^2$ where σ_ν is the Maxwell pressure, ε_0 is the vacuum permittivity, ε_r is the dielectric constant of the polymer, V is voltage, and Y_E is the modulus of elasticity. A Neo–Hookean model can be determined in the following way:

$$\sigma_m = \frac{Y_E}{3}\left(\lambda - \lambda^{-2}\right)$$

where λ is the stretch or $\lambda = (L/L_0)$, where L and L_0 are the current and initial thicknesses of the DEA that determine the change in length from 0–4000 V.

References

1. W. C. Röntgen, Ueber die durch Electricität bewirkten Form—und Volumenänderungen von dielectrischen Körpern, *Ann. Phys.*, 1880, **247**(13), 771–786.
2. R. Pelrine, R. D. Kornbluh and J. P. Joseph, Electrostriction of polymer dielectrics with compliant electrodes as a means of actuation, *Sens. Actuators, A*, 1998, **64**(1), 77–85.
3. R. Pelrine, R. Kornbluh, Q. Pei and J. Joseph, High-speed electrically actuated elastomers with strain greater than 100%, *Science*, 2000, **287**(5454), 836–839.
4. R. Pelrine, R. D. Kornbluh, Q. Pei, J. Eckerle, P. Jeuck, S. Oh and S. Stanford, Dielectric elastomer: generator mode fundamentals and applications, *Proc. SPIE-Int. Soc. Opt. Eng.*, 2001, **4329**, 148.

5. R. Pelrine, R. D. Kornbluh, Q. Pei, S. Stanford, S. Oh, J. Eckerle, R. J. Full, M. S. Rosenthal and K. Meijer, Dielectric elastomer artificial muscle actuators: toward biomimetic motion, *Proc. SPIE-Int. Soc. Opt. Eng.*, 2002, **4695**, 126.

6. G. Kofod and R. Kornbluh, Dielectric Elastomer Actuators as Intelligent Materials for Actuation, Sensing and Generation, ed. M. Shahinpoor and J. Schneider, *Intelligent Materials*, Royal Society of Chemistry, Cambridge, UK, ch. 16, 2008, pp. 396–423.

7. S. Wax and R. Sands, Electroactive polymer actuators and devices, *Proc. SPIE-Int. Soc. Opt. Eng.*, 1999, **3669**, 2.

8. L. Treloar, *The Physics of Rubber Elasticity*, Clarendon Press, Oxford, MA, 1975.

9. L. Hoss and R. J. Marczak, *A New Constitutive Model for Rubber-Like Materials, Mecánica Computacional*, ed. M. G. Eduardo Dvorkin, and M. Storti, Buenos Aires, Argentina, 15–18 November 2010, vol. XXIX, pp. 2759–2773.

10. F. Carpi, P. Chiarelli, A. Mazzoldi and D. de Rossi, Electromechanical characterisation of dielectric elastomer planar actuators: comparative evaluation of different electrode materials and different counterloads, *Sens. Actuators, A*, 2003, **107**, 85–95.

11. G. Kofod, PhD Thesis, Dielectric Elastomer Actuators, Risø National Laboratory/Danish Technical University (Pitney Bowes Management Services Denmark A/S), 2001.

12. P. A. L. S. Martins, R. M. Natal Jorge and A. J. M. Ferreira, A Comparative Study of Several Material Models for Prediction of Hyper-Elastic Properties: Application to Silicone-Rubber and Soft Tissues, *Strain*, 2006, **42**, 135–147.

13. C. Keplinger, M. Kaltenbrunner, N. Arnold and S. Bauer, Röntgen's electrode-free elastomer actuators without electromechanical pull-in instability, *Proc. Natl. Acad. Sci. U. S. A.*, 2010, **107**(10), 4505–4510.

14. L. R. G. Treloar, Stress-strain data for vulcanized rubber under various types of deformation, *Trans. Faraday Soc.*, 1944, **40**, 59–70.

15. R. S. Rivlin and D. W. Saunders, Large elastic deformations of isotropic materials VII. Experiments on the deformation of rubber, *Philos. Trans. R. Soc., A*, 1951, **243**, 251–288.

16. F. Carpi, D. De Rossi, R. Kornbluh, R. Pelrine and E. Sommer-Larsen, *Dielectric Elastomers as Electromechanical Transducers. Fundamentals, Materials, Devices, Models & Applications of an Emerging Electroactive Polymer Technology*, Elsevier, 2008.

17. F. Carpi and D. D. Rossi, Dielectric elastomer cylindrical actuators: electromechanical modeling and experimental evaluation, *Mater. Sci. Eng. C*, 2004, **24**, 555–562.

18. F. Carpi, A. Migliore, G. Serra and D. Rossi, Helical dielectric elastomer actuators, *Smart Mater. Struct.*, 2005, **14**, 1210–1216.

12 Review of Shape Memory Alloys (SMAs) as Smart Materials

Mohsen Shahinpoor

Mechanical Engineering Dept., University of Maine, USA
Email: shah@maine.edu

12.1 Introduction

The shape memory effect (SME) is a property of materials that are capable of solid phase transformation from a body-centered tetragonal form called thermoelastic martensite to a face-centered cubic superelastic called austenite. These materials are named shape–memory materials (SMMs) and their thermal versions called shape memory alloys (SMAs). This solid phase transformation from the body-centered (BCC) tetragonal martensite crystalline structure to a thermoelastic face-centered cubic (FCC) austenite crystalline phase by either temperature, stress or strain, is called a SME. These martensitic crystalline structures are capable of returning to their original shape in the austenite phase, after a large plastic deformation in the martensitic phase and return to their original shape when heated towards austenitic transformation. These novel effects are called thermal shape memory and superelasticity (elastic shape memory), respectively.[1,2]

Fundamentals of Smart Materials
Edited by Mohsen Shahinpoor
© The Royal Society of Chemistry 2020
Published by the Royal Society of Chemistry, www.rsc.org

SMAs are classed as smart materials because of their ability to undergo large deformations and to regain the original shape, either during unloading (*superelastic effect, SE*) or through a thermoelastic cycling (*SME*). As mentioned before, this is due to a solid crystalline phase (FCC to BCC) transformation (solid-to-solid) (*martensitic transformation*). This transformation also enables the SMAs to transform into a higher state symmetry crystal lattice (austenite) to a phase with a less symmetric lattice (*martensite*).[3]

12.2   Shape Memory Effect (SME)

The SME can be used to generate motion and force, while superelasticity allows energy storage. Recent successes are mainly in medical applications utilizing the superelasticity and biocompatibility of Ni–Ti alloys. The materials with SMEs belong to the family of SMMs consisting of both thermal and magnetic shape memory. Thermal shape memory alloys are called SMAs.

The shape memory characteristics of SMAs were first observed in 1932 by Chang and Read,[4] who noted the reversibility of an Au–Cd alloy, including shape recovery, as well as the observation of changes in resistivity. Greninger and Mooradian[5] observed the shape memory effect in Cu–Zn and Cu–Sn alloys in 1938.

The breakthrough in SMAs occurred in 1962, in the US Naval Ordinance Laboratory, where they were discovered by Buehler and co-workers.[6] They determined the shape memory effect in a 50% nickel–titanium (Ni–Ti), and they named the alloy Nitinol after Ni, Ti and the Naval Ordinance Laboratory. There are several alloys that exhibit thermoelastic martensitic or austenitic transformations. In certain alloys, an intermetallic phase also exists, which undergoes a solid phase transformation to martensite when cooled below a critical temperature, referred to as the Martensite start temperature or M_S. The martensitic transformation will be complete when the alloy temperature reaches the martensite finish temperature denoted by M_f.

The Nitinol alloy is then considered to be in the martensitic state, which is very ductile. If the specimen in this martensitic state is now plastically deformed, it can be restored to its original shape by heating it to the austenite start temperature A_S, where transformation to BCC austenite begins. When the temperature reaches a higher temperature, designated as the austenite finish temperature or A_f, the specimen fully recovers its initial shape before the plastic

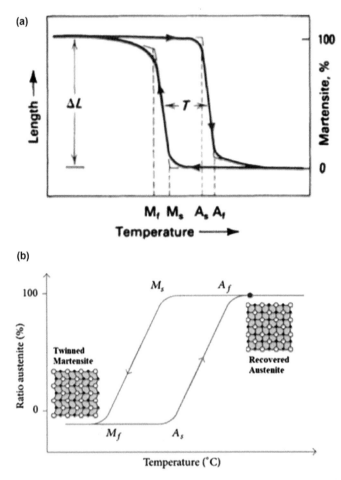

Figure 12.1 (a) Martensitic shape memory effect and (b) austenitic shape memory effect showing the twinned martensitic crystalline structure.

deformation. The critical temperatures M_S, M_f, A_S, and A_f are dependent on the SMA composition and its thermomechanical processing. The first order thermoelastic martensitic transformation shows some hysteresis (Figure 12.1a, b).

Figure 12.2 displays typical stress–strain curves for SMAs under different temperatures.

There are several SMA alloys available for various applications. Some of these SMAs are, according to ref. 1–6, Ag–Cd, Cu–Al–Ni, Cu–Sn, Cu–Zn, Cu–Zn–X (X = Si, Al, Sn), Fe–Pt, Mn–Cu, Fe–Mn–Si, Co–Ni–Al, Co–Ni–Ga, Ni–Fe–Ga, Ti–Nb, Ni–Ti, Ni–Ti–Hf, Ni–Ti–Pd, and Ni–Mn–Ga.

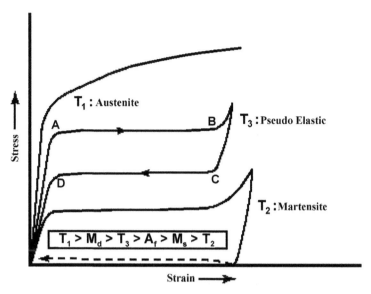

Figure 12.2 Stress–strain curves showing hysteresis for SMAs under different temperatures.

12.3 Stress–Strain–Temperature Dependence of SMAs

As described before, the SME can occur based on temperature variations or stress or strain variations, as depicted in Figure 12.3. Note that the 3D stress–strain–temperature (σ–ε–T) diagram of SMA shows the SME in the martensite state, plastic yield during the austenite/martensite phase transformation and elastic-plastic behavior of austenite at higher temperature [from ref. 13].

Note from Figure 12.3 that the unique SME property occurs due to a martensitic phase transformation between a high-symmetry cubic crystal structure or austenite crystalline phase to a low-symmetry monoclinic crystalline structure or martensite. Martensite is more stable at low temperatures and high stress, while austenite is stable at high temperatures and low stress. The mechanical behavior as a function of temperature, strain, and stress is summarized in Figure 12.3.[2] Note that below the martensite finish temperature, M_f, the SMA exhibits the SME.

Plastic deformations of martensite by stress can be recovered by increasing its temperature by direct or Joule heating the SMAs above the austenite finish temperature, A_f. At a temperature above A_f, the SMA is entirely in its austenite phase, such that upon plastic

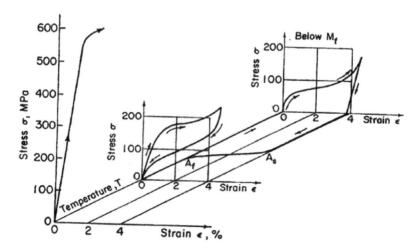

Figure 12.3 3D stress–strain–temperature (σ-ε-T) diagram of SMA showing the SME in the martensite state, plastic–elastic (pseudo elasticity, PE) deformation during austenite/martensite phase transformation and elastic–plastic behavior of austenite at higher temperature.
Reproduced from ref. 2 with permission from the Royal Society of Chemistry.

deformation by an applied load, stress-induced martensite is formed, which upon unloading however, the material returns to its austenite phase at lower stress. This behavior is called superelasticity of SMAs. The SME and superelasticity/pseudoelasticity (PE) are two distinct properties that make SMAs smart materials. The SME is a unique phenomenon by which SMAs can recover their predetermined shape by heating, even after large deformations.

Superelastic SMAs can restore their initial shape spontaneously, even from an inelastic range, upon unloading. Various compositions of SMAs such as Ni–Ti, Cu–Zn, Cu–Zn–Al, Cu–Al–Ni, Fe–Mn, Mn–Cu, Fe–Pd, and Ti–Ni–Cu have been developed, and their properties have been investigated. Among these Ni–Ti alloys is the most appropriate SMA for structural applications because of its large recoverable strain, superelasticity and exceptionally good resistance to corrosion. In this paper, unless otherwise stated, SMAs are mainly referred to as a Ni–Ti SMA (commonly known as Nitinol).

The crystal changes are also related to temperature and stress. Note the phenomena of twinning and detwinning,[3] which happen in shape memory materials as they make a solid crystalline phase transformation from austenite to twinned martensite and then accommodates to detwinned martensite, which upon heating releases any imposed stress and transforms back to austenite.

12.4 SME Variations

The annealing or heat treatment methods employed to establish memorized shapes in both shape memory and superelastic forms of Nitinol are rather similar. First one deforms an SMA sample (wire, sheet, plate, shell, *etc.*) into a desirable shape, which is constrained during the heat treatment and then places it in an oven to anneal.

The annealing parameters are usually obtained experimentally for the desired requirements. Annealing temperatures of around 500 °C and annealing times of up 30 minutes may be necessary to establish certain shapes and forms. Rapid cooling in a water quench or a rapid air cool may be desirable. The desired Nitinol shapes and properties are determined by the annealing time, maximum annealing temperature, and quenching time. Thus, the **one-way SME** is established. Depending on the manufacturing methodologies and processing techniques employed in producing the SME, SMAs can display two-way shape memory behavior as described below:

12.5 One-way SME (OWSME)

This effect is the basic mechanism of the SME. As described above this phenomenon is called the one-way SME (OWSME). The OWSME is a fundamental property of SMAs, as depicted in Figure 12.3. Macroscopically, the mechanism of the OWSME: (a) starts with martensite, then (b) loads and deforms the martensite phase $T \leq M_f$, (c) is heated above A_s (austenite), then (d) cooled to martensite $T \leq M_f$.

12.6 Two-way SME (TWSME)

It is possible to also remember the shape of the martensitic phase under certain conditions. Thus, the SMA will show two different shapes, one in heating to A_f and one in cooling to M_f. This behavior is called the two-way shape memory effect (TWSME), in contrast to the OWSME. The mechanism responsible for the two-way effect – *i.e.*, the training process introduces defects in the lattice that generate localized stress – upon cooling these stress fields causes preferred variants of martensite to form. This property enables SMAs to change shape spontaneously on both heating and cooling. It is thus possible to modify the shape of the SMAs in a reversible way between two different ones *via* the changing of temperature between A_f and M_f.

Exercise 12.1: Describe SMAs and the SME.

Answer to Exercise 12.1: SMAs belong to the large class of smart materials because of their ability to undergo large deformations and to regain their original shape, either during unloading (*superelastic effect*) or through thermoelastic cycling (*SME*). The SME originates from the fact that solid phase transformation from the body-centered tetragonal martensite crystalline structure to a thermoelastic face-centered cubic austenite crystalline phase by either temperature, stress, or strain is possible. These martensitic crystalline structures are capable of returning to their original shape in the austenite phase, after a large plastic deformation in the martensitic phase and return to their original shape when heated towards austenitic transformation, which is the SME.

12.7 Constitutive Equations for SMAs

Here, we present various popular theories in connection with stress–strain–temperature relationships in SMAs. Constitutive equations generally describe the physical behavior of SMA in terms of dealing with stresses, strains, and temperature variations, as well as their rates of change with time. Tanaka[7] presented the first plausible constitutive equations for the SMAs. Liang and Rogers[8] represented Tanaka's exponential model[7] in the trigonometric cosine form. These early models did not quite fit the relevant experimental data of Prahlad and Chopra.[9] One reason for this was the lack of inclusion of the stress-induced martensitic volume fraction at a low temperature in the constitutive equations. Another shortcoming for both of these early models[7,8] was that they only considered phase transformation from martensite to austenite and back based on temperature variations.

Since stress-induced martensite can be generated at lower temperatures, and stress-induced martensite interacts with the temperature-induced martensite, it can influence the behavior of SMAs. Thus, Brinson proposed his model,[10–12] in which the martensite volume fraction is divided into two fractions, namely the stress- and temperature-induced martensitic volume fractions. In the following section, these constitutive models are briefly described.

Exercise 12.2: What is the main difference between the Tanaka SMA constitutive equations and Liang and Rogers constitutive equations and models?

Answer to Exercise 12.2: Tanaka's model of the SME considers exponential functions for time variations of pertinent functions, while

Liang and Rogers present the trigonometric Cosine form of the functional variations in constitutive equations.

12.8 Tanaka Model

The Tanaka formulation[7] maintains that the thermomechanical behavior of the SMAs is fully expressed by three major state variables: strain, ε, temperature, T, and martensite fraction, ξ. The following formulations of these constitutive equations, Tanaka's model,[7] Liang and Rogers' model[8] and Brinson's model[10] closely follows the instructional presentation of Elahinia and Ahmadian,[14] in 2005. Figure 12.4 displays the critical stress–temperature profile used in Tanaka's constitutive equations. The constitutive equation in this model relating the state variable stress (σ), strain (ε), and temperature (T) in terms of martensitic volume fraction (ξ) is:

$$\sigma - \sigma_0 = E(\xi)(\varepsilon - \varepsilon_0) + \theta(T - T_0) + \Omega(\xi)(\xi - \xi_0) \tag{12.1}$$

where $(\sigma_0, \varepsilon_0, \xi_0, T_0)$ represent the initial state or original condition of the material. In this equation, E is the module of elasticity and assumed to be a linear function of the martensite volume fraction:

$$E(\xi) = E_A + \xi(E_M - E_A) \tag{12.2}$$

and Ω is the phase transformation coefficient and is given by eqn (12.3):

$$\Omega(\xi) = -\varepsilon_L E(\xi) \tag{12.3}$$

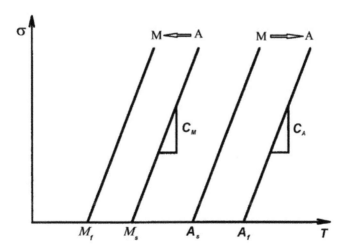

Figure 12.4 Critical stress–temperature profiles used in the Tanaka model.

where ε_L is the maximum recoverable strain or residual strain. The kinetics equations describe the martensite volume fraction as an exponential function of stress and temperature are:

$$\xi_{A \to M} = 1 - \exp(a_M(M_s - T) + b_M \sigma) \text{ for } T > M_f \text{ and } C_M(T - M_s) < \sigma < C_M(T - M_f)$$

$$(12.4)$$

Based on the experimental data, the two material constants C_A and C_M, which are called stress-influence coefficients, show the influence of stress on the transition transformations. Note that:

$$\xi_{M \to A} = \exp(a_A(A_s - T) + b_A \sigma) \text{ for } T > A_s \text{ and } C_A(T - A_f) < \sigma < C_A(T - A_s)$$

$$(12.5)$$

where a_A, a_M, b_A, and b_M are material constants in terms of the transition temperatures A_s, A_f, M_s, and M_f. Figure 12.4 depicts the two material constants C_A and C_M in relation to the slopes of the martensitic and austenitic transformations and stress temperature variations proposed in Tanaka's model.

12.9   The Liang and Roger Model[8]

This model presents constitutive equations similar to the constitutive equation that the Tanaka model proposes, except that exponential dependence is replaced by trigonometric cosine dependence with some justification. The kinetics equations that describe the martensite volume fraction as a cosine function of stress and temperature are:

$$\xi_{A \to M} = \frac{1 - \xi_A}{2} \cos\left[a_M(T - M_f) + b_M \sigma\right] + \frac{1 + \xi_A}{2} \quad \text{for } T > M_f \text{ and } C_M(T - M_s)$$
$$< \sigma < C_M(T - M_f)$$

$$(12.6)$$

$$\xi_{M \to A} = \frac{\xi_M}{2} \cos[a_A(T - A_s) + b_A \sigma) + 1] \quad \text{for } T > A_s \text{ and } C_A(T - A_f)$$

$$(12.7)$$

$$< \sigma < C_A(T - A_s)$$

where a_A, a_M, b_A, b_M are four material constants:

$$a_A = \frac{\pi}{(A_f - A_s)}, b_A = \frac{-a_A}{C_A}, a_M = \frac{\pi}{(M_s - M_f)}, b_M = \frac{-a_M}{C_M} \qquad (12.8)$$

Furthermore, ξ_M and ξ_A are the initial martensite volume fractions before the current transformation.

Both the Tanaka and Liang and Rogers models can only explain the phase transformation from martensite to austenite and its reverse transformation. However, SMAs also show such solid phase transformation under stress at low temperature. Thus, Brinson and co-researchers came up with a new theory, more consistent with experimental results, as discussed in the next section.

12.10   The Brinson Model[10-12]

Since the SME at a lower temperature is caused by the conversion between stress-induced martensite and temperature-induced martensite, these models cannot be implemented in the detwinning of martensite that is responsible for the SME.[10-12] This problem was solved by Brinson and his model.[10-12]

In this model, the martensite volume fraction (ξ) is separated into stress-induced (ξ_s) and temperature-induced martensite fractions (ξ_T):

$$\xi = \xi_s + \xi_T \tag{12.8}$$

The first form of the constitutive equation in this model was:

$$\sigma - \sigma_0 = E(\xi)\varepsilon - E(\xi_0)\varepsilon_0 + \Omega_s(\xi)\xi_s - \Omega_s(\xi_0)\xi_{s0} + \Omega_T(\xi)\xi_T - \Omega_T(\xi_0)\xi_{T0} + \Theta(T - T_0) \tag{12.9}$$

but it was shown by Brinson and Huang[11] that the constitutive eqn (12.9) could be reduced to the simplified form of:

$$\sigma - \sigma_0 = E(\xi)(\varepsilon - \varepsilon_L\xi_s) + \Theta(T - T_0) \tag{12.10}$$

where ε_L is the residual strain when $\sigma = 0$.

The variation in critical stresses with temperature for transformation consistent with the separation of ξ into two components is shown schematically in Figure 12.5.

Evolved equations for the calculation of the martensite fractions as a function of temperature and stress can now be represented in accordance with Figure 12.5, as follows:

Conversion to detwinned martensite:
For $T > M_s$ and $\sigma_s^{cr} + C_M(T - M_s) < \sigma < \sigma_f^{cr} + C_M(T - M_s)$

$$\xi_s = \frac{1 - \xi_{s0}}{2}\cos\left\{\frac{\pi}{\sigma_s^{cr} - \sigma_f^{cr}}\left[\sigma - \sigma_f^{cr} - C_M(T - M_s)\right]\right\}$$

$$+ \frac{1 + \xi_{s0}}{2}, \xi_T = \xi_{T0} - \frac{\xi_{T0}}{1 - \xi_{s0}}(\xi_s - \xi_{s0}) \tag{12.11}$$

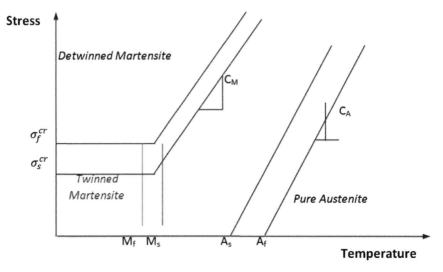

Figure 12.5 Critical stress–temperature profiles used in the Brinson model.

For $T < M_s$ and $\sigma_s^{cr} < \sigma < \sigma_f^{cr}$

$$\xi_s = \frac{1 - \xi_{s0}}{2} \cos\left[\frac{\pi}{\sigma_s^{cr} - \sigma_f^{cr}}(\sigma - \sigma_f^{cr})\right]$$

$$+ \frac{1 + \xi_{s0}}{2}, \xi_T = \xi_{T0} - \frac{\xi_{T0}}{1 - \xi_{s0}}(\xi_s - \xi_{s0}) + \Delta_{T\varepsilon}$$

(12.12)

where, if $M_f < T < M_s$ and $T < T_0$ then $\Delta_{T\varepsilon} = \dfrac{1 - \xi_{T0}}{2}\{\cos[a_M(T - M_f)] + 1\}$
otherwise $\Delta_{T\varepsilon} = 0$

Conversion to Austenite:
For $T > A_s$ and $C_A(T - A_f) < \sigma < C_A(T - A_s)$

$$\xi = \frac{\xi_0}{2}\left\{\cos\left[a_A\left(T - A_s - \frac{\sigma}{C_A}\right)\right] + 1\right\}, \xi_s = \xi_{s0}$$

(12.13)

$$- \frac{\xi_{s0}}{\xi_0}(\xi_0 - \xi), \xi_T = \xi_{T0} - \frac{\xi_{T0}}{\xi_0}(\xi_0 - \xi)$$

This completes our formulation of constitutive equations for SMAs.

12.11 Cardiovascular Superelastic Stents

According to ref. 2, stents (endoprostheses) are medical implants for dilating and maintaining the proper opening of a

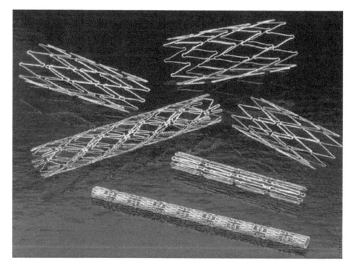

Figure 12.6 Laser cut tubular stents.
Reproduced from ref. 2 with permission from the Royal Society of Chemistry.

previously occluded anatomic passage such as a vein or artery (Figure 12.6).

A superelastic stent, which has been collapsed and then inserted through a catheter to the occluded region of the artery, at the coronary artery location, and when pushed out of the confining sheath expands to push back the plaque and make contact with and support the artery wall.

In all cases, the NiTi used is in the superelastic condition. Typical laser cut stents are illustrated in Figure 12.6.

12.12 Medical Applications

There are also many medical applications of SMAs, such as in orthodontic and dental procedures, superelastic medical devices, and most important of all, as cardiovascular stents to open up blocked arteries due to the accumulation of plaque on their internal walls. Refer to the research work of Shahinpoor and Wang[15–19] for some general considerations on large motion SMA actuators.

Refer to studies by Wang and Shahinpoor[20] for the applications of rotatory SMA actuators to knee and leg muscle exercisers by Ferrara *et al.*[21] for applications of trained SMA fixtures in spinal instrumentation for the correction of spinal deformities, to Moneim *et al.*[22] for flexor tendon repair using SMA suture.

12.13   SMA Engineering and Industrial Applications

There are many applications connected to SMMs. These include couplings, seals, electrical connectors, virtual two-way actuators, non-biased safety devices, thermal interrupters, eyeglass frames, cellular phone antennas, and home appliances. There are also many scientific experiments that utilize SMMs as an experimental means to perform other experiments.

Adaptive structures, structural damping, high force devices, jet engines and other aeronautical applications, including the active backend rotors of helicopters. Refer to work by Shahinpoor[23] for fibrous parallel spring-loaded SMA robotic linear actuators (Figure 12.7), and to the study by Shahinpoor and Martinez[24] for the applications of SMAs as thaw sensors.

12.14   Conclusions

The SME was introduced in this chapter as a property of materials capable of solid phase transformation from a body-centered tetragonal form called thermoelastic martensite to a face-centered cubic superelastic phase called austenite. These materials are named SMMs and the thermal version of those called SMAs. It was described how the martensitic crystalline structures are capable of returning to their original shape in the austenite phase, after a large plastic deformation in the martensitic phase and returning to their original shape when heated to achieve austenitic transformation.

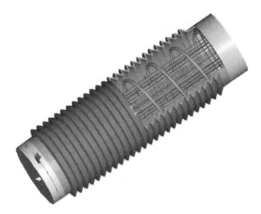

Figure 12.7   Fibrous parallel spring-loaded SMA robotic linear actuator with resilience.

These novel effects are called thermal shape memory and super-elasticity (elastic shape memory), respectively. As mentioned before this is due to a solid crystalline phase (FCC to BCC) transformation (solid-to-solid) (*martensitic transformation*). This transformation also enables SMAs to transform to a higher state symmetry crystal lattice (austenite) to a phase with a less symmetric lattice (*martensite*). Three major models of SMA behavior, attributed to Tanaka, Liang and Rogers, and Brinson, were also analyzed and compared. The major differences and similarities between these models have also been elaborated upon and presented in this chapter based on experimental data of the shape memory and superelastic behavior of Ni–Ti SMA wires.

Homework Problems

Homework Problem 12.1

Describe the design of a parallel spring-loaded fibrous SMA actuators, as depicted in Figure 12.7.

Homework Problem 12.2

Describe how SMAs can be used as thaw sensors.

Homework Problem 12.3

What are the differences between the Tanaka and Brinson model of SMAs?

Homework Problem 12.4

Describe Brinson's constitutive equations for SMAs.

References

1. L. McDonald Schetky, Shape Memory Alloys, *Sci. Am.*, 1979, **241**(5), 74–82.
2. L. McDonald Schetky, Shape Memory Alloys as Multi-Functional Materials, in *Intelligent Materials*, ed. Shahinpoor and Schneider, Royal Society of Chemistry, Cambridge, ch. 13, 2008, pp. 317–333.
3. D. C. Lagoudlas, *Shape Memory Alloys: Modeling and Engineering Applications*, Springer Publishing, London, New York, 2008.
4. L. C. Chang and T. A. Read, The Gold-Cadmium Beta Phase, *Trans. AIME*, 1951, **191**, 47–52.
5. A. B. Greninger and V. G. Mooradian, Strain transformation in metastable beta copper-zinc and beta copper-tin alloys, *Trans. AIME*, 1938, **128**, 337–368.

6. W. J. Beuhler, J. V. Gilfrich and R. C. Wiley, Effect of low-temperature phase changes on the mechanical properties of alloys near composition TiNi, *J. Appl. Phys.*, 1963, **34**, 1467–1477.
7. K. Tanaka, A thermomechanical sketch of shape memory effect: one-dimensional tensile behavior, *Res. Mechanica.*, 1986, **18**, 251–263.
8. C. Liang and C. A. Rogers, One-dimensional thermomechanical constitutive relations for shape memory material, *J. Intell. Mater. Syst. Struct.*, 1990, **1**, 207–234.
9. H. Prahlad and I. Chopra, Comparative Evaluation of Shape Memory Alloy Constitutive Models with Experimental Data, *J. Intell. Mater. Syst. Struct.*, 2001, **12**, 383–395.
10. L. C. Brinson, One-dimensional constitutive behavior of shape memory alloys: thermomechanical derivation with non-constant material functions, *J. Intell. Mater. Syst. Struct.*, 1993, **4**, 229–242.
11. L. Brinson and M. Huang, Simplifications and comparisons of shape memory alloy constitutive models, *J. Intell. Mater. Syst. Struct.*, 1996, 7, 108–114.
12. A. Bekker and L. C. Brinson, Phase diagram based description of the hysteresis behavior of shape memory alloys, *Acta Mater.*, 1998, **46**, 3649–3655.
13. J. G. Boyd and D. C. Lagoudas, A thermodynamic constitutive model for the shape memory materials part I. The monolithic shape memory alloys, *Int. J. Plast.*, 1998, **6**, 805–842.
14. M. H. Elahinia and M. Ahmadian, An enhanced SMA phenomenological model: I. The shortcomings of the existing models, *J. Smart Mater. Struct.*, 2005, **14**, 1297–1308.
15. M. Shahinpoor and G. Wang, Electro-Thermo-Mechanics of SMA Fiber Bundles Embedded in An Elastic Medium, *Recent Advances in Engineering Science, Proc. 31st Technical Conference of the Society of Engineering Science, Symposium on Active Materials and Smart Structures*, Texas A&M University, College Station Texas, October 1994, pp. 408–409.
16. G. Wang and M. Shahinpoor, A New Design for A Rotatory Joint Actuator Made with Shape Memory Alloy Contractile Wire, *Int. J. Intell. Mater. Syst. Struct.*, 1997, **8**(3), 191–279.
17. G. Wang and M. Shahinpoor, Design, Prototyping and Computer Simulation of a Novel Large Bending Actuator Made with a Shape Memory Alloy Contractile Wire, *Smart Mater. Struct. Int. J.*, 1997, **6**(2), 214–221.
18. G. Wang and M. Shahinpoor, Design for Shape Memory Alloy Rotatory Joint Actuators Using Shape memory Effect and Pseudoelastic Effect, *Smart Mater. Technol.*, 1997, **3040**, 23–30.
19. G. Wang, A General Design of Bias Force Shape Memory Alloy (BFSMA) Actuators and An Electrically-Controlled SMA Knee and Leg Muscle Exerciser for Paraplegics and Quadriplegics, PhD Thesis, Department of Mechanical Engineering, University of New Mexico, Albuquerque, New Mexico, 1998.
20. G. Wang and M. Shahinpoor, Design of A Knee and Leg Muscle Exerciser Using A Shape memory Alloy Rotary Actuator, *Proc. SPIE Smart Materials and Structures Conference*, March 3–5, 1998, San Diego, California, Publication No. SPIE 3324-29, 1998.
21. L. Ferrara, M. Jaeger, A. Keshavarzi, M. Shahinpoor and E. Benzel, The Use of Trained Shape Memory Alloy Fixtures in Spinal Instrumentation for the Correction of Spinal Deformities, Presented at the Congress of Neurological Surgeons Annual Meeting (poster 368), Boston, MA, November 1999.
22. M. S. Moneim, K. Firoozbakhsh, A. A. Mustapha, K. Larsen and M. Shahinpoor, Flexor Tendon Repair Using Shape Memory Alloy Suture, *J. Clin. Orthop. Relat. Res.*, 2002, **402**, 251–259.
23. M. Shahinpoor, Fibrous, Parallel Spring-Loaded Shape-Memory Alloy (SMA) Robotic Linear Actuators, *U. S. Pat.* 5,821,664, 1998.
24. M. Shahinpoor and D. R. Martinez, Shape Memory Alloy Thaw Sensors, *U. S. Pat. Off.*, 5,735,607, 1998.

13 Review of Magnetic Shape Memory Smart Materials

Mohsen Shahinpoor

Mechanical Engineering Dept., University of Maine, USA
Email: shah@maine.edu

13.1 Introduction

Magnetic shape memory alloys (MSMAs) or materials (MSMMs), also often referred to as ferromagnetic shape memory alloys (FSMAs), are an interesting member of the family of shape memory materials (SMMs). FSMAs combine the attributes and properties of ferromagnetism with reversible martensitic–austenitic crystalline solid phase transformations that feature twining. Magnetically-controlled shape memory (MSM) materials represent a new way to produce motion and force.

MSM phenomena were originally observed in 1996 by Ullakko[1] and O'Handley, Ullakko, Huang, and Kantner[2] and O'Handley,[3] when they demonstrated the magnetic shape memory effect for a Ni–Mn–Ga alloy. A new mechanism was proposed[1–3] based on the magnetic field-induced reorientation of the twin microstructures of a MSMA similar to a shape memory alloy (SMA). It was observed[1–3] that the MSM effect shows a much more rapid response in terms of actuation compared with the rather slow response of thermal SMAs. Here, the magnetic field controls the reorientation of the twin variants in the MSMA analogous to the way that they are controlled by stress in classical thermal SMAs.

Fundamentals of Smart Materials
Edited by Mohsen Shahinpoor
© The Royal Society of Chemistry 2020
Published by the Royal Society of Chemistry, www.rsc.org

The MSM effect shows that certain SMMs that are ferromagnetic can also show very large dimensional changes up to 10% under the application of an external magnetic field. These strains occur within the low-temperature (martensitic) phase.

According to Webster and Ziebeck,[4] MSMAs are a family of materials in the larger family of Heusler materials. These are ordered inter-metallic materials with the generic formula X_2YZ, where X and Y are 3D elements and Z is a group IIIA–VA element, as shown in ref. 4. Ferro-magnetic shape memory (FSM) materials combine the properties of ferromagnetism with those of a reversible martensitic transformation. These materials, such as the MSMA Ni_2MnGa with a cubic Heusler structure in the high-temperature austenitic phase is capable of undergoing a cubic-to-tetragonal martensitic transformation to exhibit MSM effects. Note that similar to SMAs with twinning and detwinning, when they are subjected to one externally applied magnetic field, the twins will have a favorable orientation and grow at the expense of other twins. This observation causes the shape change of MSMAs.

Compared to SMAs such as Nitinol, in which temperature is the main driving force in addition to stress and strain, FSMAs transform back to their original configuration when exposed to a specific magnetic field for the intended shape memory.

13.2 MSMA Actuators

MSMAs generally have a very fast response time of less than 0.2 ms and can operate at high frequencies and large mechanical stroke. Hysteresis issues in connection with MSM actuation are also of interest in discussing MSM actuation. Hysteresis in MSM actuators occurs between the strain and the actuator input current as well as between the strain and stress. It is clear that hysteresis generates losses in the MSM material and makes the control of some positioning system applications fairly complicated. On the other hand, due to its energy dissipation characteristics, hysteresis increases the vibrational damping capability of MSM materials. It has been shown[1–3] that the strain of MSM actuators depends on the load it has to work against.

Exercise 13.1: What causes hysteresis in MSMAs?

Answer to Exercise 13.1: There are two types of hysteresis in MSM actuators. One is due to variations in the strain *versus* the actuator input current, and the other is due to variations in the strain with stress. Hysteresis is further caused by the internal characteristics of the MSM material in the sense that it increases vibrational damping of the MSM material and prevents vibrations and overshooting of the MSM element.

13.3 Sensing and Multi-functional Properties of MSMMs

Tellina *et al.*,[5] Suorsa *et al.*,[6] Suorsa,[7] Lecce,[8] and Fahler[9] have demonstrated the actuation, sensing and voltage generation properties of MSM materials. The inverse of the MSM effect is what enables MSMMs to serve as sensors. Here, any shape change in MSMMs with non-zero net magnetization dynamically alters the imposed magnetic field and that signal can serve as a sensing signal. Thus, MSMMs possess two distinct properties of actuation and sensing. MSM applications have been used to design linear motors and velocity sensors.[1–3,10]

The microstructural reorientation of the martensitic solid phase of MSMMs generates macroscopically observable field-induced strains. Since the martensite variants have different preferred magnetization directions, applied magnetic fields can be used to generate macroscopic shape change. However, FSMAs transform back to their original configuration when exposed to a magnetic field or a resilient force. Since the application of the field is instantaneous, martensitic phase transformation can rapidly occur. MSMMs exhibit large strain (~10%) under the application of an external magnetic field.

13.4 Typical MSMA Materials

The following are the most common MSMAs:[1–3] CoNiGa, CuAlNi, FePd, FePt, LaSrCuO4, and the most studied are various forms of NiMnGa such as Ni_2MnGa. In normal Fe or Ni ferromagnets, the strains associated with the magnetostriction are in the order of 1ess than 4%, whereas materials with exceptionally large magnetostriction, for example, Tb–Dy–Fe alloys (Terfenol-D), show strains in the order of 0.1%.[1–3]

13.5 Manufacturing of MSMAs

3D printing and additive manufacturing have been two of the easiest and best ways to manufacture MSMs. Essentially, directed energy deposition of the most common MSMAs such as Ni–Mn–Ga, and additive layering of ferromagnetic martensite at room temperature will create magnetic shape memory. Post-deposition annealing of successively laminated and deposited layers and the interaction of deposition boundaries with a twin microstructure will yield the desired MSMAs. Single-crystalline MSMA foils can also be manufactured by cutting thin sheets from the single MSM martensite crystal. MSMA films can also be manufactured *via* sputtering deposition.

A sputtering manufacturing methodology for MSM polycrystalline Ni–Mn–Ga films has been reported by Ohtsuka *et al.*,[11,12] Film deposition has also been performed on polyvinyl alcohol (PVA) substrates. Ohtsuka *et al.*,[11] also discussed the consequence of heat treatment on polycrystalline Ni–Mn–Ga films in terms of their performance.

13.6 MSM Mechanism

It must be noted that MSMAs are essentially different from traditional magnetostrictive materials, such as Terfenol-D and Galfenol, as they produce much larger strains of up to 10%, under relatively small imposed magnetic fields.[1–3] Magnetic anisotropy of MSM materials plays a main role in their mechanisms. The MSMA effect also requires a special microstructure that is provided by a martensitic transformation. The following Table 13.1 shows the characteristics of MSMAs according to ref. 1–3.

In addition to generating large strains of up to 10%, the MSMAs also provide innovative applications at frequencies not achievable by conventional SMAs. These advantages inspired the search for better materials such as paramagnetic and metamagnetic materials.[1–3]

There are essentially two mechanisms involved in the MSM effect and the capability to produce a large magnetic field-induced strain of typically 6%, but this can be increased up to 10%. Magnetic field-induced structural reorientation of twins and their boundary motion is one of the mechanisms of shape change. This mechanism moves the twin boundaries and makes the angles between the magnetization axis and magnetic field axis or orientation smaller. The second MSM magnetic shape memory mechanism is due to a magnetic field-induced phase

Table 13.1 Some characteristics of MSMMs.

Maximum linear deformation	100 000 $\mu\varepsilon$
Available work ($\sigma\varepsilon$)	300 MPa. 1000 $\mu\varepsilon$
Young modulus	7.7 GPa
Average resistivity	80 $\mu\Omega$ cm
Curie temperature	from 320–380 K
Change of martensitic transition temperature upon a composition change of 1%	50 K

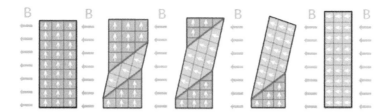

Figure 13.1 MSM working principle. Magnetically-induced reorientation of cellular domains (MIR).
Reproduced from https://commons.wikimedia.org/w/index.php?curid=51466979 under the terms of the CC BY Share Alike 3.0, Germany, licence.

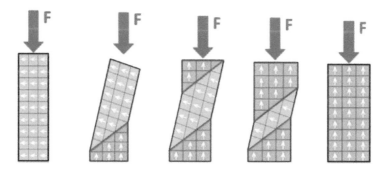

Figure 13.2 MSM working principle. Magnetically-induced Martensitic transformation (MIM).
Reproduced from https://commons.wikimedia.org/w/index.php?curid=51466979 under the terms of the CC BY Share Alike 3.0, Germany, licence.

transformation (a martensitic/austenitic phase transformation similar to that of SMAs). Figures 13.1 and 13.2 depict the MSM working principle.

The deformation kink shown in Figure 13.2 is exaggerated to illustrate the mechanism. The actual kinks are generally less than 4°. These figures are created based on a figure presented by Rene Schnetzler, and are also discussed in ref. 9, 10.

13.7 Magneto–Mechanical Constitutive Modeling of MSMs

MSMAs are essentially ferroelastic in a martensitic crystalline solid phase that possesses a redistribution of different twin variant fractions of martensite under external stress applied through the motion of twin boundaries. Ferromagnetic SMMs have an additional characteristic in terms of the magnetic activation of the deformation process in the twinned martensitic state. Based on fundamental thermodynamics principles, one can model the mechanical and magnetic characteristics and behavior of similar types of materials using the following equations:

$$\sigma = \sigma(\varepsilon, H) \tag{13.1}$$

where σ is the stress and H is the magnetic field strength. Eqn (13.1) governs the mechanical behavior of MSMMs through a nonlinear stress–strain–magnetic field equation in the presence of the magnetic field H. The magnetization value m as a function of the magnetic field H and strain ε is given by eqn (13.2):

$$m = m(\varepsilon, H) \tag{13.2}$$

Eqn (13.1) and (13.2) can be derived from the basic thermodynamic potentials and the specific Gibbs free energy $G(\epsilon, H)$ of MSMMs such that:

$$\sigma = \sigma(\varepsilon, H) = \frac{\partial \tilde{G}(\epsilon, H)}{\partial \varepsilon} \tag{13.3}$$

$$m = m(\varepsilon, H) = -\frac{\partial \tilde{G}(\epsilon, H)}{\partial \varepsilon} \tag{13.4}$$

where, $\tilde{G}(\epsilon, H) = G(\epsilon, H) - Hm(\epsilon, H)$ and $G(\epsilon, H)$ is the specific Gibbs free energy at a given temperature and pressure. Poisson's rule should be satisfied with these equations, and thus giving:[1-3]

$$\frac{\partial \sigma(\epsilon, H)}{\partial h} = -\frac{\partial m(\epsilon, H)}{\partial \varepsilon} \tag{13.5}$$

If we integrate the above equation over the range of the magnetic field H, the following equation is obtained:

$$\sigma = \sigma_0(\epsilon) - \frac{\partial \int_0^H m(\epsilon, H) dH}{\partial \varepsilon} \tag{13.6}$$

Here, $\sigma(\epsilon, 0) = \sigma_0(\epsilon)$ represents the pure mechanical stress resulting at $H = 0$. The additional magnetic field induced stress is represented by the second term on the right-hand side of eqn (13.6). In a particularly important case when $\sigma = 0$, eqn (13.6) yields the following important relationship:

$$\sigma_0(\epsilon) = \frac{\partial \int_0^h dhm(\epsilon, h)}{\partial \varepsilon} \tag{13.7}$$

One can derive the linearized version of the above equation in the form represented in eqn (13.8):

$$\in^{MSM}(h) = \left(\frac{d\sigma_0}{d\in}\right)^{-1}_{\in\,=\,0} \left(\frac{\partial \int_0^h dhm(\in, h)}{\partial \varepsilon}\right)_{\varepsilon\,=\,0} \tag{13.8}$$

According to the above equations, the magnetization and its dependence on the strain are the reasons for the MSM effect.

Exercise Number 13.2: Discuss the differences between the mechanically induced stress and magnetically induced stress in MSMMs.

Answer to Exercise No. 13.2: The pure mechanical stress occurs when the magnetic field $H = 0$ such that $\sigma(\in, 0) = \sigma_0(\in)$, resulting from the mechanical deformation of the material at $H = 0$. The additional magnetic field induced stress is represented by the second term on the right in eqn (13.6), or

$$\sigma_H = -\frac{\partial \int_0^h dhm(\in, h)}{\partial \varepsilon} \tag{13.9}$$

13.8 Some Additional Applications of MSMs

In addition to **actuators** and **sensor applications**,[1–3] the MSM effect can be used to design and develop other MSMM-based systems. Of course, in relation to actuators, the MSMM element changes its dimensions when placed in an external magnetic field. Similar to SMAs, the deformations of MSMMs in a magnetic field are fully reversible, either *via* a resilient mechanism such as a spring or by applying a magnetic field at 90° to the original field. The shape change is very rapid, and cycle times of 1–2 kHz have been observed,[1–3] and several hundred million cycles of loading and unloading have been achieved while testing fatigue life.[1–3] Note that MSMMs give much higher strain outputs (up to 10%) and higher energy densities. The following applications are also common:

13.8.1 Breaker Switch/Fuse

This application allows removes the magnetic field by temperature-induced expansion of a probe, for safety reasons. The actuator runs until the safe working temperature is reached, after that it extends further, cutting the magnetic field generation.

13.8.2 Energy Harvesters

Variable magnetic permeability under varying stress can be used to "harvest" vibration energy. Possible uses include battery charging in environments where it is difficult to gain access to new batteries for replacement.

13.8.3 Vibration Dampers

MSMMs can be used as vibration dampers by absorbing energy *via* energy harvesting.

13.8.4 Precision Positioning

Ferromagnetic SMAs can be used to develop devices and systems for the high precision positioning of nano-objects with an accuracy of down to 20 nm. The devices do not consume energy after being put in place and have applications that range from medical implants to mirror positioning in high-power telescopes.

13.9 Conclusions

This chapter introduced a new class of smart materials that belong to the general family of SMMs. MSMAs or MSMMs, often also referred to as FSMAs, combine the attributes and properties of ferromagnetism with reversible martensitic–austenitic crystalline solid phase transformations involving twinning. Magnetically controlled MSMs present a new way to produce motion and force. The chapter further presented some modeling of constitutive equations and the notion of mechanical stress, as well as magnetically-induced stresses in such materials. It further described the magnetic field-induced reorientation of the twin microstructures of a MSMA similar to that observed in SMAs. It has been reported that the MSM effect shows a much more rapid response in terms of actuation compared with the rather slow response of thermal SMAs. Here, the magnetic field controls the reorientation of the twin variants in MSMAs analogous to the twin variants being controlled by stress in classical thermal SMAs.

Homework Problems

Homework Problem 13.1

Present a comparison between MSMAs, thermal shape memory alloys, magnetostrictive materials, and piezoelectric materials.

Homework Problem 13.2

Describe the properties of magnetic shape memory linear actuators using Ni–Mn–Ga actuating elements.

References

1. K. Ullakko, Magnetically Controlled Shape Memory Alloys: A New Class of Actuator Materials, *J. Mater. Eng. Perform.*, 1996, **5**, 405–409.
2. R. C. O'Handley, K. Ullakko, J. K. Huang and C. Kantner, Large Magnetic-field-induced Strains in Ni_2MnGa Single Crystals, *Appl. Phys. Lett.*, 1996, **69**, 1966–1971.
3. R. C. O'Handley, Model for strain and magnetostriction in magnetic shape memory alloys, *J. Appl. Phys.*, 1998, **83**, 3263–3271.
4. P. J. Webster and K. R. A. Ziebeck, Heusler alloys, in *Alloys and Compounds of d-Elements with Main Group Elements. Part 2. Landolt-Börnstein—Group III Condensed Matter*, Springer-Verlag, Berlin, Germany, 1988, pp. 75–79.
5. J. Tellinen, I. Suorsa, A. Jääskeläinen, I. Aaltio and K. Ullakko, Basic Properties of Magnetic Shape Memory Actuators, Proceedings of the 8th international conference ACTUATOR 2002, Bremen, Germany, 10–12 June 2002.
6. I. Suorsa, J. Tellinen, K. Ullakko and E. Pagounis, *J. Appl. Phys.*, 2004, **95**(12), 8054–8058.
7. I. Suorsa, Performance and Modeling of Magnetic Shape Memory Actuators and Sensors, PhD Thesis, Department of Electrical and Communications Engineering, Helsinki University of Technology, Espoo, Finland, 2005.
8. L. Lecce, Shape Memory Alloy Engineering: For Aerospace, *Structural and Biomedical Applications*, Elsevier, London, 2014.
9. S. Faehler, An introduction to actuation mechanisms of Magnetic Shape Memory Alloys, *ECS Trans.*, 2007, **3**(25), 155–163.
10. M. Kohl, M. Gueltig, V. Pinneker, Y. Ruizhi, F. Wendler and B. Krevet, Magnetic Shape Memory Microactuators, *Micromachines*, 2014, **5**, 1135–1160.
11. M. Ohtsuka and K. Itagaki, Effect of heat treatment on properties of Ni–Mn–Ga films prepared by a sputtering method, *Int. J. Appl. Electromagn. Mech.*, 2000, **12**, 49–59.
12. M. Ohtsuka, Y. Konno, M. Matsumoto, T. Takagi and K. Itagaki, Magnetic-field-induced two-way shape memory effect of ferromagnetic Ni2MnGa sputtered films, *Mater. Trans.*, 2006, **47**, 625–630.

14 Shape Memory Polymers (SMPs) as Smart Materials

Mohsen Shahinpoor

Mechanical Engineering Dept., University of Maine, USA
Email: shah@maine.edu

14.1 Introduction

Shape memory polymers (SMPs) belong to the family of shape memory materials (SMM), which can be deformed into a predetermined shape under some the effect of temperature, and electric or magnetic fields, as well as strain and stress. These shapes can be relaxed back to their original field-free shapes under thermal, electrical, magnetic, strain, stress, temperature, laser, or environmental stimuli. These transformations are essentially due to the elastic energy stored in SMMs during their initial deformation.[1–3] As a member of the SMM family, SMPs are stimuli-sensitive polymers, which require the use of either heat or laser light energy to stimulate a change in shape.

The thermally-induced shape memory effect can be observed by irradiation with infrared light, exposure to alternating magnetic fields or an electric field, or immersion in water. Upon forming the SMP into its permanent shape by injection molding or extruding, it can then be deformed into another temporary shape. The initial permanent shape can then be recovered from the temporary shape upon the application of an external stimulus. This process of permanent shape recovery can be repeated several times with different imposed shape changes or deformation. Compared to shape memory alloys,

Fundamentals of Smart Materials
Edited by Mohsen Shahinpoor
© The Royal Society of Chemistry 2020
Published by the Royal Society of Chemistry, www.rsc.org

SMPs allow larger deformations between temporary and permanent shapes. SMP properties can be adjusted for specific applications such as actuators, smart fabrics,[4,5] heat-shrinkable tubes for electronics or films for packaging,[6] sun sails in spacecrafts,[7] self-disassembling mobile phones,[8] intelligent medical devices[9] and implants for minimally invasive surgery.[10,11] SMPs are generally composed of polymer networks containing molecular switches, which can be actuated by external stimuli.[4,12] In these networks, the crosslinking chain segments determine the permanent shape. These crosslinks are either covalent bonds, or they are physical in the sense that the crosslinks consist of intermolecular interactions such as van der Waals interactions or hydrogen bonds. The SMP polymer network displays shape memory functionality if the material can be temporarily deformed. Reversibly covalent crosslinking for fixing the temporary shape can be achieved *via* the attachment of functional groups to the chain segments. It has been established[1,2] that the introduction of such functional group capability to undergo photoreversible reactions has made it possible to use light as a possible stimulus for shape recovery. Other stimuli like electrical current or alternating magnetic fields are used for indirect heating of the material to create the thermally-induced shape memory effect. In this chapter, developments in the field of thermally-induced shape memory polymers are presented.

14.2 Shape Memory Polymers (SMPs) in Temperature Fields

The thermally-induced polymeric shape memory effect results from the deformation of the material in a desired operating temperature range relevant for the desired medical or industrial applications. Figure 14.1[1-4] depicts the essential mechanism of shape memory in phase-segregated linear block copolymers in which one phase acts as a cross-link determining the permanent shape and the other phase (the network chain) as the molecular switch[12] because their flexibility is a function of temperature, in particular the thermal transition temperature, T_{trans}. Above the thermal transition T_{trans}, the chain segments are flexible. However, below the T_{trans}, the polymer network is less flexible. Certain recoiling of switching segments of the polymer molecular network is driven by entropy increase due to the recoiling of elastic segments.[1,2,4]

Figure 14.2 depicts a typical SMP stress–strain–temperature curve, taken from a study by Karger-Kocsis and Kéki.[3]

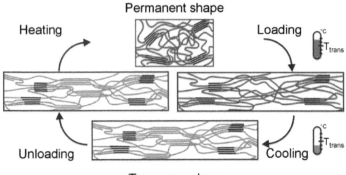

Figure 14.1   Molecular reorientation mechanisms of the thermally-induced shape memory effects in a linear block copolymer with T_{trans} = thermal transition temperature related to the switching phase. Reproduced from ref. 1 with permission from the Royal Society of Chemistry.

According to Behl, Langer, and Lendlein,[1,2] two important quantities describe shape memory effects, the strain recovery rate (R_r) and strain fixity rate (R_f). The strain recovery rate describes the ability of the material to memorize its permanent shape. The strain fixity rate R_f describes the ability to switch segments to fix the mechanical deformation. According to,[1,2]

$$R(N) = [\varepsilon_m - \varepsilon_p(N)]/[\varepsilon_m - \varepsilon_p(N-1)] \qquad (14.1)$$

$$R_f(N) = \varepsilon_p(N)/\varepsilon_m \qquad (14.2)$$

where N is the number of cycles, ε_m is the maximum strain imposed on the material, and $\varepsilon_p(N)$ and $\varepsilon_p(N-1)$ are the strains of the sample in two successive cycles in the stress-free state. The following equations, according to[1,2] describe the shape memory effect in terms of R_f and R_f:

$$R_f(N) = 1 - E_r/E_g \qquad (14.3)$$

$$R_r(N) = 1 - \{e_{vf}/[e_\alpha(1 - E_r/E_g)]\} \qquad (14.4)$$

where E_r is the rubbery modulus, E_g is the glassy modulus, e_{vf} is the viscous flow strain, and e_α is the strain for $T > T_g$.

Exercise 14.1 What is the essential mechanism of shape memory?

Answer to Exercise 14.1: Phase-segregated linear block copolymers possess one phase which acts as a cross-link determining the

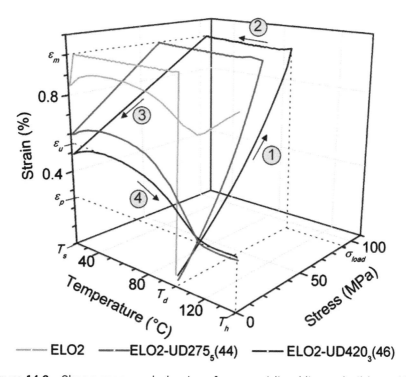

— — — ELO2   — — — ELO2-UD275$_5$(44)   — — — ELO2-UD420$_3$(46)

Figure 14.2 Shape memory behavior of an epoxidized linseed oil-based bio resin (ELO2, $T_g \approx 90$ °C) and its UD-flax reinforced composites. The numbers indicate the steps of the shape memory experiment, first (1), deforming at $T=100$ °C to $\varepsilon_m = 1\%$, second (2), shape fixing occurs by cooling to a lower $T=20$ °C while keeping $\varepsilon_m = 1\%$, third (3), unloading, fourth (4), stress-free shape recovery.
Reproduced from ref. 3, https://doi.org/10.3390/polym 10010034, under the terms of a CC BY 4.0 license, https://creativecommons.org/licenses/by/4.0/.

permanent shape and possesses another phase (the network chain) as a molecular switch. The resulting interaction produces the shape memory effect. In other words, two important quantities describe shape memory effects, the strain recovery rate (R_r) and strain fixity rate R_f. The strain recovery rate describes the ability of the material to memorize its permanent shape. The strain fixity rate, R_f, describes the ability to switch segments to fix the mechanical deformation.[1,2]

Exercise 14.2 What is the relationship between R_f and R_f?

Answer to Exercise 14.2:

From eqn (14.3) and (14.4) one can relate R_f to R_f using the ratio $E_r/E_g = 1 - R_f(N)$, Such that

$$R(N) = 1 - \{e_{vf}/[e_\alpha R_f (N)]\} \tag{14.5}$$

14.3   Thermoplastic SMPs

According to ref. 4, polyurethanes, polyether–ester, and ABA triblock copolymers are linear block copolymers with $T_{\text{trans}} = T_{\text{m}}$, where the ABA triblock is a copolymer of polytetrahydrofuran (B block) and poly(2-methyl-2-oxazoline) (A block). In polyurethanes containing poly(ε-caprolactone) a switching segment ($T_{\text{m}} = 38$–$59\ ^\circ$C) can be created.

The crystalline domains of these thermoplastic polymers work as physical crosslinkers resulting in elastic behavior above T_{g} as the switching temperature is given by the glass transition ($T_{\text{trans}} = T_{\text{g}}$).

14.4   SMP Product Development

Cornerstone Research Group (CRG), located in Dayton, OH, USA, has developed a deployable wing using a Veritex shape memory composite as a matrix. According to Hiltz,[13] a poly(norbornene)-based polymer, was first reported by the CDF Chimie Company, France, in 1984 and was made commercially available in the same year by Nippon Zeon Company of Japan under the trade name Norsorex. This polymer has a T_{g} value of between 35–40 $^\circ$C. In 1987, the Kurare Corporation of Japan also developed a second commercial poly(*trans*-isoprene)-based SMP called Kurare TP-301. Another SMP by the name of Asmer was introduced by the Asahi Company, Japan, and is poly(styrene–butadiene) based with a T_{g} value in the range of 60–90 $^\circ$C. Diisocyanate/polyol-based polyurethane SMPs were developed by Mitsubishi Heavy Industries in the late 1980s. Polyurethane-based SMPs are available under several trade names, including Diary MM-4510, which is a polyester polyol-based polyurethane. The advantage of polyurethane SMPs is their flexibility. Also, these polyurethane SMPs are thermoplastic. Nippon Zeon has produced a series of polyester-based SMPs that are marketed under the trade name Shable. The thermoplastic nature of polyurethanes renders them amenable to processing *via* techniques such as extrusion, injection or blow molding or by solution casting.

14.5   Thermomechanical Constitutive Equations for SMPs

There are several papers on the constitutive modeling of shape memory polymers by Hu *et al.*,[14] Zhang and Yang,[15] Nguyen,[16] Liu *et al.*,[17]

Kim *et al.*,[18] and Tobuschi *et al.*[19,20] Here, we present a summary of the equations presented in these studies. For a particular shape to be memorized by an SMP, a load should be applied to the polymer at a constant higher temperature followed by cooling the SMP to a lower temperature at constant strain. Residual stress builds up within the SMP during cooling. The stress is then removed from the SMP at a constant low temperature in the third step. Finally, the SMP is heated to a high temperature in the fourth step, and the stored strain is released, where after the SMP returns to the initial shape. These four steps to determine the kinematics and energy balance of the SMPs are shown below:

1. First, pre-loading the SMPs
2. Second, constrained shape memorization
3. Third, constraint release
4. Fourth, unconstrained strain storage release

The SMP is described as a composite material where two phases (glassy and rubbery) coexist in the temperature range that define the glass transition.[17-20] The volume fraction of the glassy phase depends on temperature only.[17-20] The glassy (or "frozen") phase, with a volume fraction $f_f(T)$, has elastic behavior that stems from internal energy, which arises from interatomic van der Waals interactions and covalent intramolecular interactions. On the other hand, the elastic behavior of the rubbery (or "active") phase, with a volume fraction of $f_g(T)$, which is equal to $(1 - f_f(T))$, strongly depends on the entropy of the macromolecular chains. According to ref. 14, in a cooling process, the glass transition develops from a high-temperature, T, equal or greater than $T_g = T_h$, where $f_f(T) = 0$ to a low-temperature T smaller than or equal to T_l, where $f_f(T) = 1$. The $f_f(T)$ function plays a central role in the model and has the following empirical form based on constitutive models presented in ref. 17–20:

$$f_f(T) = 1 - 1/[1 + c(T_h - T)^n] \tag{14.6}$$

where c is a constant, which gives $f_f(T_h) = 0$ but not $f_f(T_l) = 1$. Therefore, the following empirical relationship, which generates similar curves,

$$f_f(T) = [1 - (T - T_l)/(T_h - T_l)^m]^n \tag{14.7}$$

is preferred here, taking $f_f(T) = 0$ if $T > T_h$ and $f_f(T) = 1$ if $T < T_l$. The integers m and n are just arbitrary.

To develop a constitutive model for describing the thermo–mechanical behavior of such thermo-responsive SMPs mathematically, two-phase models consisting of rubbery and glassy phases have been used, *e.g.*, a parallel connection between the rubbery and glassy phases (Kim *et al.*[18]). At high T_h, the rubbery phase is the dominant micro-structure, while the glassy phase is dominant at low T_l. The switching between these two phases is modeled by the volume fractions of each phase, *e.g.*, at T_h, the volume fractions of the rubbery and glassy phases are 1 and 0, respectively. We followed this methodology to derive the constitutive equation for SMPs.

Exercise 14.3 What are the steps needed to determine the kinematics and energy balance of SMPs?

Answer to Exercise 14.3: There are four steps to determine the kinematics and energy balance of the SMPs, first, pre-loading of the SMPs, second, constrained shape memorization, third, constraint release, and fourth, unconstrained strain storage release

Three types of modeling exist for SMPs:

1. Modeling based on storage deformation
2. Modeling based on phase transition
3. Modeling based on viscoelasticity

At a given temperature T, the total Helmholtz energy of SMP is H_t, which is a sum of the Helmholtz energy of the rubbery phase, H_r, plus the Helmholtz energy of the glassy phase, H_g, such that:

$$H_t = f_r(T) H_r + f_g(T) H_g \tag{14.8}$$

where $f_r(T) = f_a(T)$ and $f_g(T) = f_f(T)$ are volume fractions of the rubbery (active phase) and glassy phase (frozen phase), respectively, such that $f_r(T) + f_g(T) = 1$. Furthermore, simple constitutive equations in terms of the total stress, s_t, and total strain, e, of the SMP can be postulated to be given in terms of stress in the rubbery or active phase, $s_r = s_a$, and stress in the glassy or frozen phase, $s_g = s_f$, such that:

$$s_t = f_r(T) s_r + f_g(T) s_g = f_a(T)s_a + f_f(T)s_f = (1 - f_f(T))s_a + f_f(T)s_f \tag{14.9}$$

Similarly, in terms of the total strain and corresponding strain of the rubbery phase, e_a, and the glassy phase, e_f:

$$e_t = f_r(T) e_r + f_g(T) e_g = f_a(T)e_a + f_f(T)e_f = (1 - f_f(T)) e_a + f_f(T) e_f \tag{14.10}$$

Furthermore, the volume fraction of the frozen or glassy phase is given by:

$$f_f(T) = 1 - \{1/[1 + C_f(T_h - T)^n]\}. \tag{14.11}$$

where T_h is a reference temperature, and C_f and n are fitting parameters determined by appropriate experiments. Tobushi *et al.*[19] published a constitutive model for the shape memory polymer thermomechanical cycle. They added a slip mechanism due to internal friction based on the standard linear viscoelastic model responsible for the irreversible strain. In this early model, the threshold strain was calculated using:

$$\dot{\varepsilon} = \dot{\sigma}/E + \sigma/\mu - (\varepsilon - \varepsilon_s)/\lambda + \alpha T \tag{14.12}$$

In the above equation, the slip strain is $e_s = C(e_{creep} - e_0)$, where e_0 is the threshold (reference) strain. A nonlinear constitutive model was later proposed by Tobushi *et al.*,[20] in their improved version of the previous model.[19] Additional nonlinear elastic and nonlinear viscoelastic stress terms were added to the previous constitutive equation. We refrain from reproducing these equations and instead refer the reader to ref. 20. The resulting constitutive model is given by eqn (14.13):

$$\dot{\varepsilon} = \frac{\dot{\sigma}}{E} + m\left\{\frac{(\sigma - \sigma_y)}{k}\right\}^{(m-1)}\left(\frac{\dot{\sigma}}{k}\right) + \frac{\sigma}{\mu} + \left(\frac{1}{b}\right)\left\{\left(\frac{\sigma}{\sigma_c}\right) - 1\right\}^n - \frac{(\varepsilon - \varepsilon_s)}{\lambda} + \alpha T \cdot \tag{14.13}$$

where s_y is the yield stress of the glassy phase, s_c is the critical stress, E is the Young's modulus of the SMP, m is the viscosity of the molten phase, a is the coefficient of thermal expansion, and m, n, k, b and l are obtained from experimental procedures. Based on the thermodynamic concepts of entropy and internal energy, it is possible to interpret the thermo mechanical behavior of SMPs from a macroscopic viewpoint without explicitly incorporating details of the molecular interactions. In 2017, Yunxin[21] presented a sophisticated model and thorough 3D constitutive equations for SMPs, which we refrain from covering here due to space limitations.

14.6 Conclusion and Outlook

This chapter introduced the family of SMPs. It also described the polymeric shape memory effect as an interaction between the rubbery and glassy phases, induced by temperature and deformation fields (stress/strain). Two important quantities are used to describe the shape memory effect; the strain recovery rate (R_r) and strain fixity rate (R_f). The strain recovery rate describes the ability of the material to memorize its permanent shape. The strain fixity rate R_f describes the ability to switch segments to fix the mechanical deformation.

Constitutive thermomechanical equations were also presented and discussed in terms of the shape memory effect in SMPs. Potential applications were briefly discussed, which include biomedical implants, bio switches, biosensors, smart fabrics, intelligent packaging, self-healing, and self-repairing automobiles. Other application areas include the aerospace industry and reusable composite tooling.

Homework Problems

Homework Problem 14.1

Discuss the types of modeling of SMPs.

Homework Problem 14.2

How is the total strain, e_t, related to the strains of the rubbery phase, e_f, and the strain of the glassy phase, e_g?

References

1. M. Behl, R. Langer and A. Lendlein, Intelligent Materials: Shape Memory Alloys, in *Intelligent Materials*, ed. M. Shahinpoor and H. J. Schneider, Royal Society of Chemistry, Cambridge, UK, 2008.
2. Shape-Memory Polymers, *Advances in Polymers Science*, ed. A. Lendlein, Springer Publishers, Berlin/Heidelberg, vol. 226, 2010.
3. J. Karger-Kocsis and S. Kéki, Review of Progress in Shape Memory Epoxies and Their Composites-Revie, *Polymers*, 2018, **10**(1), 34–44.
4. M. Ken Gall, N. A. Mikulas, F. Munshi, Beavers and M. Tupper, Carbon Fiber Reinforced Shape Memory Polymer Composites, *J. Intell. Mater. Struct.*, 2000, **11**, 877.
5. G. J. Monkman, Advances in shape memory polymer actuation, *Mechatronics*, 2000, **10**, 489.
6. K. Otsuka and C. M. Wayman, *Shape Memory Materials*, Cambridge University Press, Cambridge, UK, 1988.
7. H. Tobushi, S. Hayashi, A. Ikai and H. Hara, Thermomechanical Properties of Shape Memory Polymers of Polyurethane Series and their Applications, *J. De Phys. IV, Colloque C1*, 1996, **6**, C1–377.

8. K. Kitagawa, M. Moritoki and N. Nishiguchi, What is High Pressure Crystalization Process, *CEER, Chem. Econ. Eng. Rev.*, 1984, **16**(12), 30–35.
9. C. Liang, C. A. Rogers and E. Malafeew, Investigation of Shape Memory Polymers and Their Hybrid Composites, *J. Intell. Mater. Syst. Struct.*, 1997, **8**, 380.
10. K. Nakayama, Properties and Applications of Shape Memory Polymers, *Int. Polym. Sci. Technol.*, 1991, **18**, 43.
11. Y. Shirai and S. Hayashi, Development of Polymeric Shape Memory Material, MTB184, Mitsubishi Heavy Industries, Inc., Japan, 1988.
12. Z. G. Wei and R. Sandstrom, Review Shape-memory materials and hybrid composites for smart materials, *J. Mater. Sci.*, 1998, **33**, 3743.
13. J. A. Hiltz, Shape Memory Polymers: Literature Review, Defence Research and Development Canada, Technical Memorandum DRDC Atlantic TM 2002-127, August 2002.
14. J. Hu, Y. Zhu, H. Huang and J. Lu, Recent advances in shape memory polymers: structure, mechanism, functionality, modeling, and application, *Prog. Polym. Sci.*, 2012, **37**, 1720–1763.
15. T. D. Nguyen, Modeling Shape-Memory Behavior of Polymers, *Polym. Rev.*, 2013, **53**, 130–152.
16. Q. Zhang and Q. S. Yang, Recent advance on constitutive models of thermal-sensitive shape memory polymers, *J. Appl. Polym. Sci.*, 2011, **123**(3), 1502–1508.
17. Y. Liu, K. Gall, M. L. Dunn, A. R. Greenberg and J. Diani, Thermomechanics of shape memory polymers: uniaxial experiments and constitutive modeling, *Int. J. Plast.*, 2006, **22**, 279–313.
18. J. H. Kim, T. J. Kang and W.-R. Yu, Thermo-mechanical constitutive modeling of shape memory polyurethanes using a phenomenological approach, *Int. J. Plast.*, 2010, **26**, 204–218.
19. H. Tobushi, T. Hashimoto and S. Hayashi, Thermomechanical Constitutive Modeling in Shape Memory Polymer of Polyurethane Series, *J. Intell. Mater. Syst. Struct.*, 1997, **8**(8), 711–718.
20. H. Tobushi, K. Okumura, S. Hayashi and N. Ito, Thermomechanical constitutive model of shape memory polymer, *Mech. Mater.*, 2001, **33**(10), 545–554.
21. Y. Li, Y. He and Z. Liu, A viscoelastic constitutive model for shape memory polymers based on multiplicative decompositions of the deformation gradient, *Int. J. Plast.*, 2017, **91**, 317.

15 Review of Smart Materials for Controlled Drug Release

Carmen Alvarez-Lorenzo* and Angel Concheiro

Departamento de Farmacología, Farmacia y Tecnología Farmacéutica,
R + DPharma Group (GI-1645), Facultad de Farmacia and Health Research Institute
of Santiago de Compostela (IDIS), Universidade de Santiago de Compostela, 15782
Santiago de Compostela, Spain
*Email: carmen.alvarez.lorenzo@usc.es

15.1 Introduction

Systemically-administered controlled release systems allow fine tuning of drug bioavailability, by regulating the amount and the rate at which the drug reaches the bloodstream, which is critical for the success of the therapy. Some drugs pose important problems in terms of efficacy and safety (*e.g.*, antitumor drugs, antimicrobials) and suffer instability problems in the biological environment (*e.g.*, gene material), and thus the therapeutic performance of these drugs is improved when they are selectively directed (targeted) from the bloodstream to the site of action (tissues, cells or cellular structures). Both macro-dosage forms and nano-delivery systems may notably benefit from stimuli-responsive materials. Differently to pre-programmed drug release systems, formulations that provide discontinuous release as a function of specific signals (stimuli) are advantageous in many situations. Triggering drug release where, when, and how it is needed requires detailed knowledge of the changes that the illness causes in physiological parameters. These changes can be characterized in terms of biomarkers

Fundamentals of Smart Materials
Edited by Mohsen Shahinpoor
© The Royal Society of Chemistry 2020
Published by the Royal Society of Chemistry, www.rsc.org

(*e.g.*, glucose, specific enzymes, or *quorum sensing* signals in the case of infection) and physicochemical parameters (pH, ions, temperature, glutathione) that may be exploited as internal stimuli. When the physio-pathological changes are too weak or poorly specific, the application of external stimuli may be an alternative. External sources of temperature, ultrasound, light, and magnetic or electric fields may allow for the focal switch on/off of drug release. This chapter provides an overview of the interest of activation-modulated and feedback-regulated controlled release systems, the mechanisms behind them, and some specific examples of responsive materials and their applications.

15.2 Drug Dosage Forms and Drug Delivery Systems

The active pharmaceutical ingredient (the drug) of a medicine is seldom directly administered to the body. In most cases, the drug is mixed with some inert substances (named excipients) and after some processing is transformed into a medicine (*i.e.*, a drug dosage form) that exhibits suitable properties to be administered through a certain route. Transforming the drug into a drug dosage form is a quite complex process that is carefully conducted observing strict quality control rules to ensure efficacy and safety.

The incorporation of drugs into dosage forms allows us to solve problems of different complexity and to respond to very diverse needs. For example, when the required dose is very small – a few milligrams or even less than one milligram – the dosage form gives it a sufficient mass and an adequate structure so that patients or health personnel can easily handle it. The dosage form provides the drug with a physical state (solid, liquid, or aerosol) appropriate for facile dosing and administration, and also for good acceptance by the patients. In the specific case of oral administration, the correction of the unpleasant taste of many drugs is a very important problem for treatment adherence, which can be solved by designing the dosage form correctly.

From a technological point of view, the dosage form can play important roles, such as correcting the physicochemical and biopharmaceutical properties of the drug; namely solubility or permeability, which are critical for the effectiveness of the treatments. Most pharmacological treatments require that the drug has access to the bloodstream, from where it can reach the site of action. The drug can be directly injected into the blood using intravascular (abbreviated as *i.v.*) dosage forms. Differently, if an extravasal route (*e.g.*, oral) is used, the drug has to overcome various biological barriers before it reaches the bloodstream. When the drugs are poorly soluble in water, complete dissolution of the dose (a mandatory first step) is difficult. This is a problem that affects up to 40% of drug candidates. In other cases, the hydrophilicity or the ionic character of the drug is very marked, or its molecular weight is very large, which hinders the passage of the dissolved drug molecules through biological membranes; namely, the drug exhibits low permeability. The incorporation of the drug into a well-designed dosage form can effectively contribute towards overcoming these problems.[1,2]

The dosage form decisively affects the bioavailability, that is, the fraction of dose and the rate at which the drug reaches the bloodstream, which are determinants for the evolution of plasma drug levels. Once the distribution equilibrium is reached, the concentration of drug in the therapeutic target site is, in general, proportional to the concentration in the plasma.

As a consequence, the efficacy of most treatments is directly related to the evolution of plasma levels over time. It is possible to establish for each drug a certain therapeutic margin, that is to say, a range of concentrations that provokes in most patients an adequate pharmacological effect, without manifesting serious side effects. Below the lower limit (minimum effective concentration) there are no therapeutic effects, and if the upper limit (minimum toxic concentration) is exceeded, the risk of relevant side effects is unacceptable. Thus, *conventional dosage forms rely on that the access of the drug to its site of action is produced by "flooding" the organism, which implies the indiscriminate exposure of all organs and tissues.* This is the way in which most medicines still work, and this approach is adequate for most drugs.

Nevertheless, it is also true that some drugs pose important problems in terms of efficacy and safety (*e.g.*, antitumor drugs, antimicrobials) and suffer instability problems in the biological environment (*e.g.*, gene materials) when systemically administered. The therapeutic performance of these kinds of drugs can be notably improved if they are selectively directed from the bloodstream to the site of action, accumulating in the specific organs, tissues, cells or cellular structures where the therapeutic effect is required. This selective addressing of the drug to certain structures is known as *targeting* and can be accomplished by incorporating the drug molecules into nanometre size carriers.[3] These nanocarriers can cross the biological barriers through a variety of pathways and direct the drug through passive or active mechanisms to the place of interest, where the release should finally take place. *Since the performance of the targeted nanocarriers resembles the role of a courier that delivers the package at the right place, they are also called drug delivery systems.* Drug delivery systems should not release the drug into the bloodstream, so a correlation between plasma concentration and concentration in the target environment cannot be expected.

15.3 Interest of Smart Materials for Controlled Drug Release

Both macro-dosage forms and nano-delivery systems can notably benefit from stimuli-responsive materials, especially when precise control of the drug release process is required. Differently to the acute (emergency) cases that demand rapid access of the drug to the bloodstream and thus a fast drug release from the dosage form is mandatory, many treatments are intended to be prolonged for several days, weeks or even years for addressing chronic pathologies. The *first generation of controlled release systems* (started in the 1970s) was envisioned as platforms (tablets, implants, *etc.*) able to prolong drug release as much as possible according to a *pre-programmed rate*. The aim was that these drug dosage forms release the drug in a predictable way and disregard the status of the patient to maintain the drug levels in the therapeutic margin for a while, minimizing the number of intakes. This approach notably improves the efficacy of treatments with short half-life drugs using lower dose per day, which also helps to minimize adverse events and facilitates patient compliance. To meet the demands for prolonged release, novel excipients appeared on the scene, with the commitment of regulating the release by dissolution, diffusion, erosion, or osmotic mechanisms.[4] With the exception of osmotic systems that require osmotic-active substances of low molecular weight, the other mechanisms rely on conformational changes that polymers may undergo when coming into contact with aqueous physiological fluids; this water-sensitiveness is reflected in a decrease in the glass transition temperature and/or formation of gels which regulate drug release.

The *second generation of controlled release systems* aimed to regulate not only the release duration but also the moment at which the release should be triggered.[1] This *activation-modulated drug release* was first conceived to protect labile drugs from the harsh environment of the stomach and to prevent side effects in the regions where the drug is not intended to act or to be absorbed.[5] The release is activated by some physical, chemical, or biochemical processes. For example, polymers with pH-dependent solubility, time-dependent swelling, or sensitivity to enzyme degradation in certain regions of the gastro-intestinal tract are widely used for this purpose.[6,7] Responsive oral dosage forms can also be designed for the site-specific treatment of colon pathologies, such as inflammatory bowel disease or colon cancer, by selective release of the drug in this region, which maximizes the contact of the whole drug dose with the affected mucosa.[8]

The *third-generation of controlled release systems* were intended to perform as *feedback-regulated drug release systems*. The purpose is to regulate drug release to match it to the physio/pathological conditions of the body, particularly to the progression of the illness.[9-12] To accomplish this challenging aim, the system (either a macro-dosage form or a nano-drug delivery system) should simultaneously perform as a sensor for a certain biomarker and as an actuator to switch the release on and off. The drug release rate should be regulated as a function of the intensity/concentration of a triggering agent, such as a biochemical substance that may serve as an index of the pathological state. When the triggering agent is above a certain level, the release is activated. This activation induces a decrease in the level of the triggering agent and, finally, the drug release is stopped. Thus, advanced excipients should now act as sensors and actuators, imitating the recognition role of enzymes, membrane receptors, and antibodies in living organisms for the regulation of chemical reactions and maintenance of the homeostatic equilibrium.

From a therapeutic point of view, a discontinuous release as a function of specific signals (stimuli) is profitable in many situations, particularly when (i) the drug should be released only if required, for example, the glucose-dependent release of insulin, or at the time at which the symptoms are more intense, as occurs for illnesses that follow circadian rhythms (*e.g.*, hormones, and drugs for heart rhythm disorders or asthma); (ii) the drug has to be released only in the pathological region of a tissue, but not in the healthy region, such as in the case of cancer chemotherapeutics or antimicrobial therapies that require delivery systems able to identify the affected cells and selectively release the drug there, with minimal exposure to other organs or cells; and (iii) the drug has to be released only in certain compartments inside the cells to avoid premature degradation, as occurs in the case of therapeutic peptides and gene therapy.[13,14] Thus, stimuli-responsive (or smart) controlled release systems are valuable not only for new drug candidates and sophisticated biopharmaceuticals, which usually

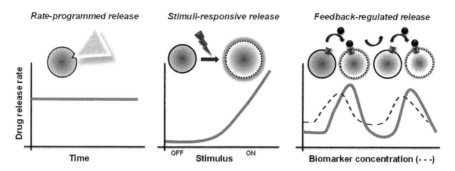

Figure 15.1 Typical drug release patterns provided by the first (rate-programmed), the second (stimuli-responsive) and the third (feedback regulated) controlled release systems.

exhibit deficient biopharmaceutical and stability properties, but they may also give added value to already-in-use drugs to fully exploit their therapeutic potential.[15–17]

Performance of the three generations of controlled release systems is schematically depicted in Figure 15.1.

15.4 Stimuli to Be Exploited and Applications

The design of both activation-modulated and feedback-regulated controlled release systems requires detailed knowledge of the changes that the illness causes in physiological parameters, not only at a systemic level, but mainly in the affected cells and tissues, and deeper insight into the different environments that the drug should cross until it reaches the target site. These changes and environments can be characterized in terms of biomarkers (*e.g.,* glucose, specific enzymes, or *quorum sensing* signals in the case of infection) and physicochemical parameters (pH, ions, temperature, glutathione) that may be exploited as internal stimuli to trigger the release. Systems with these performances demand "active" excipients able to modify their properties rapidly and in a predictive, reproducible way when the stimulus appears. In some cases, the activation of the system should lead to the release of the whole dose, as occurs, for example, with colon-specific dosage forms or nanocarriers that target the drug molecules to a tumor. Thus, the response to the stimuli causes an irreversible change in the system. In other cases, the stimulus (*e.g.,* glucose) provokes a reversible change in the dosage form or the nanocarrier that triggers the release of a certain amount of drug (*e.g.,* insulin), but when the stimulus disappears, the release is stopped, and the system remains ready for the next stimulus. These feedback regulated systems are also known as *self-regulated* systems.[18–21]

In some cases, the physiopathological changes are so tiny or poorly specific that the sensitiveness of the components is not sufficient to obtain a responsive system. The application of external stimuli may help to overcome this problem. External sources of temperature, ultrasound, light and magnetic or electric fields may allow for the focal switch on/off of drug release from nanocarriers that have previously accumulated in the target site. Externally activated systems are considered to work in *open circuit*, and they can provide pulsed drug release.[22]

Overall, smart controlled release systems require advanced excipients that proactively interact with the stimulus and undergo reversible changes in their properties, which are transmitted to the whole system in the form of dramatic changes in drug release rate. These advanced excipients are also useful for the development of theranostic systems that combine diagnosis and drug delivery capabilities in a single entity. The stimuli that can be *a priori* useful to regulate drug release are quite diverse (Figures 15.2 and 15.3).

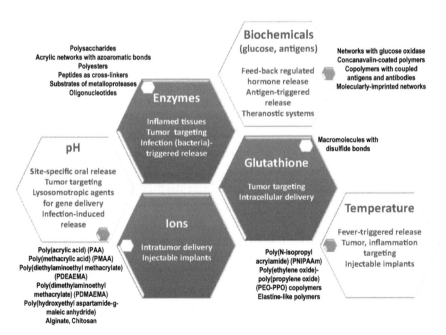

Figure 15.2 Examples of materials (outside the hexagons) responsive to internal stimuli that can be integrated into smart controlled drug release systems and some applications (inside the hexagons).

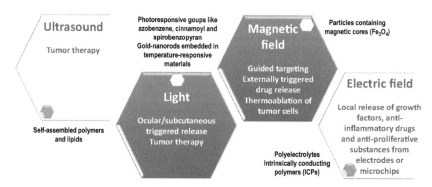

Figure 15.3 Examples of materials (outside the hexagons) responsive to external stimuli that can be integrated into smart controlled drug release systems and some applications (inside the hexagons).

They can induce phase transitions in the responsive components without altering their chemical composition (*e.g.,* assembly/disassembly, collapse/swelling) but, in some situations, they can alter chemical groups or bonds (*e.g.,* through enzymatic or redox reactions) and modify the conformation, solubility or integrity of the delivery system.[23–25,28] In the first case, when only phase transitions are involved, the changes induced by the stimulus are mostly reversible when the stimulus disappears, enabling repeated pulsate release.[4,11,25] In the second one, when the chemical groups are also altered, the delivery system is mainly conceived to avoid premature leakage of the drug until the stimulus appears and, if this has enough intensity, the complete discharge of drug may be triggered.[26–29] Thermodynamic transitions coupled to chemical transformation are common in Nature. In fact, biopolymers that can undergo phase transitions are essential for all evolved-life forms, since they are the only materials that can fulfill the three main requirements of the living systems: i) minimal complexity to form and function, ii) ability to produce different structures in a reproducible way, and iii) ability to transmit all information necessary to the forms and functions.[11]

15.4.1   pH- and/or Ion-responsive Systems

The pH gradients that exist under healthy and pathological conditions have been widely explored as a stimulus to trigger drug release. The characteristic changes in pH along the gastrointestinal tract have been largely exploited to achieve site-specific oral delivery.

Additionally, the extracellular pH of tumor tissues (6.5–7.0) is more acidic than those of the blood and the healthy tissues (7.4).[30] Inside cells, the differences in pH among the cytosol (7.4), Golgi apparatus (6.4), endosome (5.5–6.0), and lysosome (5.0) are considerable.[31] Inflamed tissues and wounds are also characterized by a decrease in pH to 5.4–7.2, and non-healing wounds can achieve relevant alkaline pH values, up to 8.9.[32] The growth of microorganisms can itself notably alter the pH of the affected tissue, but can also induce the release of bodily enzymes (*e.g.*, metalloenzymes) that cause further modifications in pH along the healing time. The pH of a wound can even be used as an index of its likelihood to completely recover.[33] Semen has also been shown to induce notable changes in vaginal pH.[34]

Lipids and polymers that behave as weak acids or bases with pK_a values in the pH range of interest may be suitable to develop pH-responsive drug release systems. Carboxylic acid, sulfonate, and primary and tertiary amino groups can modify the degree of ionization due to pH modifications in the physiological range. The change from neutral to ionized state dramatically alters the conformation and the affinity of the chains for the solvent as well as the interactions among them. The neutralization of the charges makes water become a poor solvent. Thus, an increase in the degree of ionization may be translated in the disassembly of weakly bonded components or the swelling of covalent networks. For example, networks bearing acid groups swell at alkaline pH but collapse (shrink) at low pH, while those bearing bases swell in acid medium and shrink when the pH rises. Polyampholyte systems containing both types of monomers show minimum swelling at neutral pH, *i.e.*, when both acids and bases are partially ionized, and they interact among each other.[35] It should also be noted that ionic strength in general, but certain ions in particular (especially di- or multi-valence ones) can notably affect the pH-responsiveness and also the conformation of the polymer chains, altering the affinity for water and the swelling, and even inducing self-associations that trigger sol–gel transitions.[36,37] Therefore, choosing suitable components, it is possible to develop systems that are responsive to almost any situation in the body that involves a change in pH or in the concentration of ions that can form complexes with ionizable groups.[18] Remarkable examples of pH-responsive systems are nanogels for tumor-targeting delivery, lysosomotropic micelles, and liposomes designed for gene delivery, micelles that adhere to the teeth and release antimicrobial agents only at cariogenic pH,[38] vaginal gel networks for semen-induced release, and *in situ* gelling intraocular depots.

15.4.2 Enzyme-responsive Systems

Enzymes can be exploited to break certain bonds, causing disassembly or rupture of the system and thus triggering drug release.[39] Furthermore, disregulation, namely hypo/hyperexpression, of the enzymes is often associated with disease states, and thus such disregulation could be exploited to trigger the release in the affected tissues or sites of the body.[40,41] Capthesins, plasmin, urokinase-type plasminogen activator, prostate-specific antigen, matrix metalloproteases, β-glucuronidase, and carboxylesterases, are overexpressed in certain tumors.[42]

An enzyme-responsive system requires at least an enzyme-sensitive component that is a substrate of the enzyme, and a drug that can be chemically or physically entrapped in the system. To be effective, the system has to be able to reach the enzyme, which is particularly critical when the enzymatic activity is associated with a particular tissue or the enzyme is found at higher concentrations intracellularly. The most evaluated enzymes for triggering drug release are hydrolases, which can break covalent bonds that keep together certain components or can modify certain chemical groups, altering the balance of electrostatic, hydrophobic, steric or π–π interactions, van der Waals forces or hydrogen bonding.[40–42] For example, proteases can trigger the release of drugs linked to the carrier by a peptide or when the carrier is stabilized by peptide links that are substrates of the enzyme; glycosidases can induce release from polysaccharide-based carriers; lipases can trigger drug release when they hydrolyze the phospholipid building blocks of liposomes; and certain hydrolases can control the assembly or disassembly of inorganic nanoparticles and mostly the degradation of the gatekeepers of the pores in which the drug molecules are hosted.[41] Kinases and phosphatases can be used for reversible rupture of bonds and, thus, to obtain pulsating drug release.[43]

Enzyme-responsive systems have been designed in the form of supramolecular assemblies (mainly micelles and liposomes), chemically crosslinked gels and nanocontainers, and porous silica nanoparticles with responsive gatekeepers.[40,41,44] They have been shown to be useful for specific release in inflammation sites and tumor cells, and the microorganism-triggered release of antimicrobial agents.[45–48] This latter application, still scarcely explored, is intended to avoid prophylactic and prolonged use of antimicrobials that can lead to toxic effects in the patients and also favor the apparition of resistant variants.[49] For example, the high levels of thrombin-like activity found in wounds infected with *Staphylococcus aureus* have inspired the development of conjugates of gentamicin with poly(vinyl alcohol) through a thrombin-sensitive peptide linker. The conjugate released gentamicin when it was incubated with thrombin and leucine aminopeptidase together, but not with one enzyme alone. Gentamicin was successfully released upon incubation with *S. aureus* wound fluid, strongly reducing the bacterial number in an animal model of infection.[50]

Nanocarriers degradable by bacterial enzymes have also been explored as targeting agents for anticancer drug delivery because of the selective accumulation of bacteria in tumors. As an example, the attenuated strain of *S. aureus* NCTC8325 SBY1 has preferential distribution in tumor tissues and causes a unique environment in the tumor, mimicking an infectious disease. Nanogels formed with a polyphosphoester core coated with a layer of poly(ε-caprolactone), and a shell of poly(ethylene glycol) showed a rapid release of doxorubicin in the extracellular space of the tumor due to lipases secreted by the bacteria. Lipases degraded the poly(ε-caprolactone) layer and triggered drug release, which in turn led to a marked reduction in tumor volume. These results open novel strategies for anticancer therapy using controlled bacterial infection as a stimulus.[51]

15.4.3 Biomarker-responsive Systems

Systems responsive to biomarkers try to imitate the physiological self-regulating mechanisms by integrating both molecular recognition and responsive behavior in a single structure.[52] These functionalities can be achieved as follows:

(i) a specific sensor of the biomarker (for example, an enzyme) is attached to a pH-responsive network; the reaction between the enzyme and the biomarker (substrate) results in the modification of the inner pH of the network. In the absence of the biomarker, the network does not modify its conformation, but when the biomarker concentration reaches a certain level, it reacts with the enzyme resulting in a product that modifies the local pH and, consequently, induces the network to change its degree of swelling. Once the substrate is consumed, the pH restores to the original value, and the network adopts its initial degree of swelling. The swelling/collapse cycles can be exploited to switch the release of a drug on/off. For example, glucose-responsive nanocarriers have been prepared to encapsulate glucose oxidase in PEG–poly(propylene sulfide)–PEG micelles. When glucose permeates in the micelles, glucose oxidase transforms it to gluconolactone and also generates hydrogen peroxide as a side product, which in turn oxidizes the thioethers of poly(propylene sulfide) into sulfoxides and sulfones. This oxidation makes the copolymer become hydrophilic and the micelles disassemble.[53] Glucose oxidase, catalase, and insulin have been trapped together inside poly-(2-hydroxyethyl methacrylate-*co*-N, *N*-dimethylaminoethyl methacrylate) hydrogels. Simulating *in vivo* conditions, glucose oxidase converts the glucose in gluconic acid, causing ionization of the amino groups in the copolymer and thus the swelling of the network and the release of insulin. Catalase was included to provide oxygen to the oxidation reaction. A nice correlation between glucose concentration and the release rate of insulin was found.[54] The high incidence of diabetes has notably prompted the development of glucose (enzyme-based or not) biosensors that can be implanted under the skin for continuous monitoring of blood glucose concentrations.[55,56] The biosensor can be coupled to a transdermal signal reading, in the form of a "smart tattoo," that sends a warning signal in the case of hypo- or hyperglycemia.[57]

(ii) competitive mechanism, which relies on the grafting to the network of a binding agent (lectin, antigen) for the biomarker. In the absence of the target biomarker, the binding agent interacts with a complementary component in the formulation, and the network shrinks, trapping the drug molecules. When the biomarker appears or reaches a certain concentration, it competes with the complementary component to interact with the binding agent. This process causes the rupture of the network and triggers the release of entrapped drugs.[58–62] Lectins and phenylboronic acid have been shown to be useful for interacting with glycoproteins and glycolipids.[46] Immobilizing antibodies in the drug delivery system possibly triggers antigen release. These delivery systems integrate antibodies and antigens that act as crosslinking points. This highly specific interaction is broken only when free antigens appear in the medium at a concentration sufficient enough to compete with the antigens that form part of the network.[63] A step forward in this field is the use of aptamers as components of responsive gates or valves inserted in nanocarriers.[64] Aptamers are short single-stranded oligonucleotides that can be prepared to exhibit high binding selectivity and specificity for almost any kind of molecule.[65] In the absence of the molecular stimulus, aptamers form metastable structures, which become disrupted when the stimulus is recognized.[66] For example, nanocontainers (*e.g.*, silica particles) with pores loaded with the drug of interest can be capped using a variety of gatekeepers (called apta valves when formed by aptamers) that open or close the pores according to a variety of stimuli.[67,68]

Medical devices decorated with ergosterol have been explored for the fungi-triggered release of antifungal agents.[69] A variety of antifungal agents, such as polyenes, azoles and allylamines, block fungi growth by interacting with ergosterol at fungi membranes. Thus, bio-inspired medical devices decorated with ergosterol were designed to host polyenes and to selectively release them in the presence of fungi using a competitive mechanism. Ergosterol-functionalized materials efficiently retained natamycin and nystatin (<10% released at day 14) in the absence of ergosterol. Differently, in the presence of ergosterol liposomes that mimic fungi membranes, the drug release rate was 10-to-15 fold enhanced. Ergosterol-functionalized devices that were loaded with nystatin efficiently inhibited *Candida albicans* biofilm formation.[69]

A different approach used to create recognition domains in synthetic networks is molecular imprinting technology, which allows reproduction of the small, but critical, part of the biomacromolecules responsible for the interaction with their target molecule.[70–72] This approach pursues the creation of polymer networks with cavities (artificial receptors) that are sterically and chemically complementary to the target molecule, recognizing it with high selectivity. To do that, the substance of interest is used as a template during polymerization to induce an adequate arrangement of the monomers, forming complexes with some of them at an appropriate stoichiometry in a favorable solvent (Figure 15.4).

The conformation shown in Figure 15.4 is made permanent during polymerization, and the subsequent removal of the template molecules reveals imprinted cavities, which can act as artificial receptors. The conformation of the receptors, and thus, the binding affinity of the network to the drug, can be regulated through conformational changes induced by the stimulus. The arrangement is made permanent during polymerization, and finally, the artificial receptors are revealed when the template is washed out.[73–80]

15.4.4   Glutathione-responsive Systems

Glutathione (GSH)-triggered release can be exploited to obtain intracellular specific release, namely in the cytoplasm and the nucleus.[81]

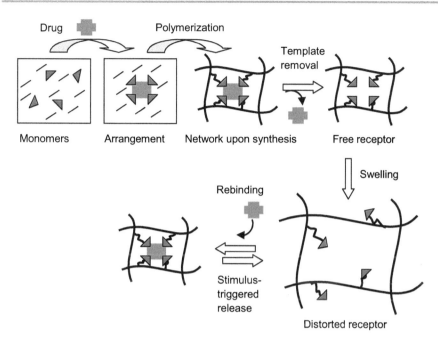

Figure 15.4 Synthesis of a molecularly imprinted network prepared with stimuli-responsive components. The drug molecules act as templates causing the arrangement of the monomers before polymerization.
Reproduced from ref. 17 with permission from Smithers Rapra, Copyright 2013.

The intracellular compartments (cytosol, mitochondria, and nucleus) contain GSH tripeptide at a concentration (2–10 mM) which is 2 to 3 orders higher than that achieved in the extracellular fluids (2–20 μM).[82] Moreover, tumor tissues have a greater concentration in GSH (4-to-7-fold higher) than the healthy ones, which reinforces the role of GSH to trigger drug release specifically in tumor cells.[83] Block copolymers, polymer networks, and crosslinking agents bearing disulfide (–S–S–) bonds are thus suitable to undergo reduction reactions in the presence of GSH, leading to the rupture of the bonds.[81] As a consequence of the redox process, the nanostructure swells or disassembles and the drug is released. The therapeutic potential of the glutathione-responsiveness has been already demonstrated using micelles and polymersomes loaded with anticancer drugs.[84] Interestingly, several antibody–drug conjugates with labile bonds responsive to the intracellular redox conditions are already in the clinical phase for personalized treatment of a variety of tumors.[85]

15.4.5 Temperature-responsive Systems

Several pathological conditions, such as infection, inflammation, infarction, or tumor processes, evolve with local increments in temperature.[86] Additionally, a localized increase in temperature can be easily achieved by applying external sources of heat on the skin or irradiating a metal-containing drug delivery system. Similarly, alternating magnetic fields can also cause a moderate increase in the temperature of the local environment of superparamagnetic particles, as will be explained in the next sections.

Temperature-responsive systems are commonly made up of polymers with a critical solubility temperature (CST), such as poly-N-isopropyl acrylamide (PNIPAAm), poly-N,N-diethylacrylamide, poly-(methyl vinyl ether) (PMVE), poly-N-vinylcaprolactam (PVCL), poly(ethylene oxide)–poly(propylene oxide) (PEO–PPO) block copolymers, certain cellulose ethers, and elastin-like polypeptides.[87] These polymers undergo reversible changes in hydrophilicity as a function of the temperature, and these changes are transmitted to the controlled release system in the form of an abrupt change in the self-association of certain components, which triggers assembly/disassembly of micelle-like or liposome-like systems or swelling/shrinking of crosslinked networks. Responsiveness in a certain temperature range can be tuned by copolymerization of the above-mentioned polymers with hydrophobic or hydrophilic monomers to tune the critical temperature down and up, respectively.

The temperature-responsive liposome formulation ThermoDox® has been demonstrated to be useful for the treatment of a variety of tumors in clinical trials.[88] ThermoDox® can passively target doxorubicin to tumor tissues and then deliver the payload in the microvasculature of the tumor when an external source of heat is applied. A local increase in temperature above 41.5 °C rapidly triggers the release of doxorubicin or other antitumor agents loaded in the liposomes, and thus the drug concentration attains therapeutic levels in a few seconds in the tumor cells, notably improving the therapeutic efficacy of the treatment.[89]

15.4.6 Ultrasound-responsive Systems

The ultrasound equipment commonly used for physiotherapy can also be applied to trigger drug release. The waves cause a local increase in temperature and bubble cavitation, facilitating the penetration of drug nanocarriers into specific regions and the triggering of drug release.[90]

Compared to other external stimuli, ultrasound can penetrate deeper into the target tissue.[91]

Polymeric micelles and delivery systems formed by layer-by-layer assemblies can be designed for drug targeting to tumors and, several hours later when a sufficient number of nanocarriers have arrived at the target cells, the application of ultrasound can reversibly destabilize the nano-carriers causing a pulsate drug release. The amount of drug released can be modulated through the control of the frequency, power density, pulse length, and inter-pulse intervals.[92–94] The *in vivo* anti-tumor effectiveness of this approach is also promoted by the cell membrane perturbation caused by ultrasound (sonoporation), which enhances the intracellular uptake of micelles, drugs, and genes.[95] Ultrasound-responsive silica-based micro and nanoparticles have been designed to exhibit a variety of performances ranging from contrast-enhanced ultrasound imaging to ablation and triggered drug release.[96]

15.4.7 Light-responsive Systems

The incident radiation of wavelength located in the 700–1400 nm interval penetrates deep into the human tissues due to the low absorption of biological molecules such as hemoglobulin, melanin, or water. Thus, this wavelength interval in the near infrared region (NIR), which is known as the biological window, can be exploited to trigger drug release in profound tissues while being innocuous.[97] Alternatively, ultraviolet light or blue light can serve as a triggering agent for topical treatments applied to the eyes, skin or mucosa, as is normally used in photodynamic therapy. Light-induced modulation of drug release, without the interference of physiological changes, can be achieved using two different approaches, as follows:

(i) photoactive groups, such as azobenzene, cynnamonyl, spirobenzopyran or triphenylmethane that undergo conformational changes at certain wavelengths.[98] For example, the *trans–cis* isomerization of the azobenzene chromophore that occurs on exposure to UV light is accompanied by an increase in the hydrophilicity, which can lead to the disassembly of polymeric micelles, liposomes or complexes with cyclodextrins. Light-induced changes in hydrophilicity have also been explored as a way to control the release of cells from biomaterial surfaces.[99]

(ii) metal particles, mainly gold nanoparticles that absorb light and transform it in local heating; if the gold particles are incorporated in a temperature-sensitive carrier, the increase in temperature can trigger drug release.[100] Many different architectures of gold-containing nanocarriers have been described for achieving anti-tumor therapy combining both light-induced hyperthermia and chemotherapy. After intravenous injection, the nanoparticles are irradiated with a fiberoptic laser causing thermal ablation of the tumor area. It should be noted that cancer cells are destroyed at temperatures (close to 43 °C) that do not affect normal cells. Moreover, the surfaces of gold nanoparticles are very suitable for the conjugation of drugs, oligonucleotides, and peptides that can be released from the nanocarrier either through light destabilization of the system or *via* photo-cleavable bonds.[101] As an example, poly(lactic-*co*-glycolic acid) (PLGA) matrix particles containing doxorubicin and covered with a gold over-layer have been shown to abruptly release doxorubicin when exposed to NIR light, causing high cancer cell toxicity, while the increase in temperature caused tissue ablation.[102] Similarly, PEGylated poly-amidoamine (PAMAM) dendrimers have been developed to integrate gold nanoparticles for photothermal therapy, with high payloads of chemotherapy agents in a hydrophobic inner space.[103] Also, the decoration of implantable medical devices with gold nanoparticles offers promise for localized treatments.[104]

15.4.8 Magnetic-responsive Systems

Magnetite (Fe_3O_4) and maghemite $(\gamma\text{-}Fe_2O_3)$ are widely used as contrast agents for magnetic resonance imaging (MRI).[105] In addition to their usefulness for diagnosis, these superparamagnetic nanoparticles can be coated with or encapsulated into polymeric structures such as micelles or placed inside microporous inorganic carriers that also contain drugs. The superparamagnetic nanoparticles make the nanocarriers movable towards a certain tissue under the action of an external magnet. Once accumulated in the tissue, the application of an alternating magnetic field causes a local increase in temperature that can be used for cell thermoablation and to trigger drug release.[106,107] Compared to other heating resources (microwaves, radio frequency, ultrasound, and lasers), magnetic hyperthermia can deliver high heat energy into deeply situated tumors without heat loss to the healthy tissues. There is already one product on the market (NanoTherm®), and several others are in the pre-clinical phase.[108,109] Commonly, the drug is released while the magnetic field is on, leading to site-specific treatment. The increase in temperature can be modulated by the frequency of oscillation and the time of application of the magnetic field; a small increase triggers reversible pulsate drug squeezing as the temperature-responsive network shrinks, while a strong increase may lead to the rupture of the carrier followed by a burst drug release and simultaneous thermal ablation of the surrounding tissues.[110,111]

15.4.9 Electricity-responsive Systems

Transdermal delivery may be facilitated, applying iontophoresis and electroporation through direct effects of the voltage on the skin. Similar equipment can be used to trigger drug release from subcutaneous dosage forms (microparticles or implants).[112] To be responsive, these dosage forms are made up of polyelectrolytes with a high density in ionizable groups. The electric field is applied through an electroconducting patch placed on the skin over the implant. The potential between the electrodes causes a movement of ions that leads to local changes in the subcutaneous pH, which in turn triggers the shrinking or the swelling of the polyelectrolyte network, modifying the drug release rate. Currently, available equipment allows for precise control of the intensity, the amount of current, the duration of the pulses, and the intervals between successive pulses. This approach has been tested for the pulsate release of insulin using subcutaneously implanted poly(-dimethylaminopropyl acrylamide) microgels,[113] and for the transdermal delivery of anti-inflammatory drugs from hydrogels of sodium alginate, carbopol, and their blends.[114,115]

Intrinsically conducting polymers (ICP), such as polypyrrole, polyaniline, or poly(3-methoxydiphenylamine), can be an alternative to polyelectrolytes.[116] ICPs are characterized by an uninterrupted and ordered π-conjugated backbone. ICPs have been used as coatings of electrodes and also as components of hydrogels. The electrical current alters the redox state of the ICP, modifying its charge and volume. The changes in the ICP structure have been explored as a way to regulate the local release of paclitaxel from stents.[117] These are growth factors from implants,[118] neurotrophins from electrodes implanted in cochlear neurons,[119] and drugs and hormones from implanted microchips,[120] or transdermal formulations.[121,122]

15.5 Conclusions and Future Aspects

Dosage forms and nanocarriers endowed with stimuli-responsive control of drug release may significantly improve the therapeutic efficiency of old and new active substances. Precise drug release at the right place, the right moment and the right rate can notably enhance the efficacy/safety ratio of most treatments using lower doses. Strong research on biocompatible materials that can respond to the small physicochemical changes that occur under certain pathological conditions is opening up an enormous number of possibilities in the design of activation-modulated and feedback-regulated controlled release systems. Although some stimuli-responsive formulations have already successfully passed preliminary clinical phases, there is still a long way to go until they are used in the clinical setting. Higher sensitiveness and reproducibility are still demanded for most applications. Inner stimuli are usually quite faint, which limits the development of feedback regulated systems. Additionally, the harsh and unsteady *in vivo* environment, which is full of many elements that can perform as disturbing signals may compromise the reproducibility of the responsiveness. In this regard, externally activatable systems have some advantages. Nevertheless, the information already gathered from *in vitro* and cell culture assays depicts a very promising future and represents solid support for the translation to animal models and clinical studies. It should be noted that stimuli-controlled release systems do not match the principles of classical pharmacokinetics in terms of absorption, distribution, and clearance. This opens up additional space for research on whether the therapeutic drug effect remains unaltered compared to conventional systemic administration (in other words, if the body response is the same) and to what extent the therapy protocols should be adapted. It can be foreseen that the information gathered under *in vivo* conditions will be highly valuable for optimizing the features of the responsive materials and the design of more efficient stimuli-responsive medicines.

Homework Problems

Homework Problem 15.1

Propose three polymers that could be useful to prepare controlled drug release systems responsive to:

a) an increase in glutathione concentration
b) matrix metalloproteases
c) the presence of a certain antigen

Homework Problem 15.2

Design an imprinted network loaded with an antimicrobial agent that exhibits self-regulated release because of pH-responsiveness.

Homework Problem 15.3

Propose a controlled drug release system that combines photothermal therapy and photo-triggered drug release.

Homework Problem 15.4

Sketch a drug release rate *versus* the application/cessation of a stimulus for (a) an activation-modulated system and (b) a feedback regulated system.

Homework Problem 15.5

Sketch the swelling behavior of (a) a pH-responsive network made of polymers with acid groups with pK_a values equal to 5.5, and (b) a temperature-responsive network made of polymers showing a lower critical solution temperature (LCST) of 34 °C. Please indicate for each case, the interval of pH or temperature where the release of a drug encapsulated in the network would take place.

References

1. A. S. Hoffman, *J. Controlled Release*, 2008, **132**, 153.
2. A. Dokoumetzidis and P. Macheras, *Int. J. Pharm.*, 2006, **321**, 1.
3. K. Park, *J. Controlled Release*, 2014, **190**, 3.
4. S. Grund, M. Bauer and D. Fischer, *Adv. Eng. Mater.*, 2011, **13**, B61.
5. I. Wilding, *Crit. Rev. Ther. Drug Carrier Syst.*, 2000, **17**, 557.
6. Y. W. Chien and S. Lin, *Clin. Pharmacokinet.*, 2002, **41**, 1267.
7. M. J. Rathbone, J. Hadgraft and M. S. Roberts, *Modified-release Drug Delivery Technology*, Marcel Dekker, New York, 2003.
8. S. Sharma and V. R. Sinha, *J. Controlled Release*, 2018, **272**, 97.
9. K. Y. Lee and S. H. Yuk, *Prog. Polym. Sci.*, 2007, **32**, 669.
10. S. S. Satav, S. Bhat and S. Thayumanavan, *Biomacromolecules*, 2010, **11**, 1735.
11. C. Alvarez-Lorenzo and A. Concheiro, *Chem. Commun.*, 2014, **50**, 7743.
12. J. Kopecek and J. Yang, *Polym. Int.*, 2007, **56**, 1078.
13. F. Levi and A. Okyar, *Expert Opin. Drug Delivery*, 2011, **8**, 1535.
14. N. I. Prasanthi, G. Swathi and S. S. Manikiran, *Int. J. Pharm. Sci. Rev. Res.*, 2011, **6**, 014.
15. W. J. Sun, Q. Y. Hu, W. Y. Ji, G. Wright and Z. Gu, *Physiol. Rev.*, 2017, **97**, 189.
16. S. Kim, J. H. Kim, O. Jeon, I. C. Kwon and K. Park, *Eur. J. Pharm. Biopharm.*, 2009, **71**, 420.
17. C. Alvarez-Lorenzo and A. Concheiro, in *Handbook of Molecularly Imprinted Polymers*, ed. C. Alvarez-Lorenzo and A. Concheiro, Smithers Rapra, Shawbury, UK, 2013, p. 309.
18. D. Schmaljohann, *Adv. Drug Delivery Rev.*, 2006, **58**, 1655.
19. F. Liu and M. W. Urban, *Prog. Polym. Sci.*, 2010, **35**, 3.
20. S. Sershen and J. West, *Adv. Drug Delivery Rev.*, 2002, **54**, 1225.
21. M. Yoshida and J. Lahann, *ACS Nano*, 2008, **2**, 1101.

22. Y. Wang and D. S. Kohane, *Nat. Rev. Mater.*, 2017, **2**, 17020.
23. M. Motornov, Y. Roiter, I. Tokarev and S. Minko, *Prog. Polym. Sci.*, 2010, **35**, 174.
24. G. Pasparakis and M. Vamvakaki, *Polym. Chem.*, 2011, **2**, 1234.
25. C. Alexander and K. M. Shakesheff, *Adv. Matter.*, 2006, **18**, 3321.
26. P. Wanakule and K. Roy, *Curr. Drug Metab.*, 2012, **13**, 42.
27. S. H. Yuk and Y. H. Bae, *Crit. Rev. Ther. Drug Carrier Syst.*, 1999, **16**, 385.
28. C. Alexander, *Expert Opin. Drug Delivery*, 2006, **3**, 573.
29. *Smart Materials for Drug Delivery*, ed. C. Alvarez-Lorenzo and A. Concheiro, Royal Society of Chemistry, London, ch. 1, 2013.
30. A. S. E. Ojugo, P. M. J. Mesheedy, D. J. O. McIntyre, C. McCoy, M. Stubbs, M. O. Leach, I. R. Judson and J. R. Griffiths, *NMR Biomed.*, 1999, **12**, 495.
31. N. Nishiyama, Y. Bae, K. Miyata, S. Fukushima and K. Kataoka, *Drug Discovery Today: Technol.*, 2005, **2**, 21.
32. G. Gethin, *Wounds UK*, 2007, **3**, 52.
33. L. A. Schneider, A. Korber, S. Grabbe and J. Dissemond, *Arch. Dermatol. Res.*, 2007, **298**, 413.
34. K. M. Gupta, S. R. Barnes, R. A. Tangaro, M. C. Roberts, D. H. Owen, D. F. Katz and P. F. Kiser, *J. Pharm. Sci.*, 2007, **96**, 670.
35. C. Alvarez-Lorenzo, H. Hiratani, K. Tanaka, K. Stancil, A. Yu Grosberg and T. Tanaka, *Langmuir*, 2001, **17**, 3616.
36. C. Alvarez-Lorenzo and A. Concheiro, *J. Controlled Release*, 2002, **80**, 247.
37. P. Mi, L. Y. Chu, X. J. Ju and C. H. Niu, *Macromol. Rapid Commun.*, 2008, **29**, 27.
38. B. Horev, M. I. Klein, G. Hwang, Y. Li, D. Kim, H. Koo and D. S. W. Benoit, *ACS Nano*, 2015, **9**, 2390.
39. S. R. Van Tomme, G. Storm and W. E. Hennink, *Int. J. Pharm.*, 2008, **355**, 1.
40. R. V. Ulijn, *J. Mater. Chem.*, 2006, **16**, 2217.
41. R. de la Rica, D. Ailia and M. M. Stevens, *Adv. Drug Delivery Rev.*, 2012, **64**, 967–978.
42. E. Fleige, M. A. Quadir and R. Haag, *Adv. Drug Delivery Rev.*, 2012, **64**, 866.
43. Z. Yang, G. Liang, L. Wang and B. Xu, *J. Am. Chem. Soc.*, 2006, **128**, 3038.
44. C. Park, H. Kim, S. Kim and C. Kim, *J. Am. Chem. Soc.*, 2009, **131**, 16614.
45. A. A. Aimetti, A. J. Machen and K. S. Anseth, *Biomaterials*, 2009, **30**, 6048.
46. T. L. Andresen, D. H. Thompson and T. Kaasgaard, *Mol. Membr. Biol.*, 2010, **7**, 353.
47. M. J. Webber, C. J. Newcomb, R. Bitton and S. I. Stupp, *Soft Matter*, 2011, **7**, 9665.
48. C. Coll, L. Mondragón, R. Martínez-Máñez, F. Sancenón, M. D. Marcos, J. Soto, P. Amorós and E. Pérez-Payá, *Angew. Chem., Int. Ed.*, 2011, **50**, 2138.
49. C. Alvarez-Lorenzo, C. A. Garcia-Gonzalez, E. Bucio and A. Concheiro, *Expert Opin. Drug Delivery*, 2016, **13**, 1109.
50. M. Tanihara, Y. Suzuki, Y. Nishimura, K. Suzuki, Y. Kakimaru and Y. Fukunisi, *J. Pharm. Sci.*, 1999, **88**, 510.
51. M. H. Xiong, Y. Bao, X. J. Du, Z. B. Tan, Q. Jiang, H. X. Wang, Y. H. Zhu and J. Wang, *ACS Nano*, 2013, **7**, 10636.
52. T. Miyata, T. Uragami and K. Nakamae, *Adv. Drug Delivery Rev.*, 2002, **54**, 79.
53. A. Napoli, M. J. Boerakker, N. Tirelli, R. J. M. Nolte, N. A. J. M. Sommerdijk and J. A. Hubbell, *Langmuir*, 2004, **20**, 3487.
54. T. Traitel, Y. Cohen and J. Kost, *Biomaterials*, 2001, **21**, 1679.
55. S. Ferri, K. Kojima and K. Sode, *J. Diabetes Sci. Technol.*, 2011, **5**, 1068.
56. R. M. DiSanto, V. Subramanian and Z. Gu, *WIREs Nanomed. Nanobiotechnol.*, 2015, **7**, 548.
57. Y. J. Heo and S. Takeuchi, *Adv. Healthcare Mater.*, 2013, **2**, 43.
58. S. Tanna, T. S. Sahota, K. Sawicka and M. J. Taylor, *Biomaterials*, 2006, **27**, 4498.
59. S. Y. Cheng, I. Constantinidis and A. Sambanis, *Biotechnol. Bioeng.*, 2006, **93**, 1079.
60. T. Miyata, N. Asami and T. Uragami, *Nature*, 1999, **399**, 766.

61. Y. Ishihara, H. S. Bazzi, V. Toader, F. Godin and H. F. Sleiman, *Chem. – Eur. J.*, 2007, **13**, 4560.
62. K. Makino, E. J. Mack, T. Okano and S. W. Kim, *J. Controlled Release*, 1990, **12**, 235.
63. K. Kataoka, H. Miyazaki, M. Bunya, T. Okano and Y. Sakurai, *J. Am. Chem. Soc.*, 1998, **120**, 12694.
64. V. C. Ozalp, F. Eyidogan and H. A. Oktem, *Pharmaceuticals*, 2011, **4**, 1137.
65. G. Mayer, *Angew. Chem., Int. Ed. Engl.*, 2009, **48**, 2672.
66. V. C. Ozalp, A. Pinto, E. Nikulina, A. Chuvilin and T. Schäfer, *Part. Part. Syst. Charact.*, 2014, **31**, 161.
67. M. Colilla, B. González and M. Vallet-Regí, *Biomater. Sci.*, 2013, **1**, 114.
68. L. Li, M. Xie, J. Wang, X. Li, C. Wang, Q. Yuan, D. W. Pang, Y. Lu and W. Tan, *Chem. Commun.*, 2013, **49**, 5823.
69. T. Segura, A. M. Puga, G. Burillo, J. Llovo, G. Brackman, T. Coenye, A. Concheiro and C. Alvarez-Lorenzo, *Biomacromolecules*, 2014, **15**, 1860.
70. C. Alvarez-Lorenzo and A. Concheiro, *J. Chromatogr. B*, 2004, **804**, 231.
71. A. Ribeiro, F. Veiga, D. Santos, J. J. Torres-Labandeira, A. Concheiro and C. Alvarez-Lorenzo, *Biomacromolecules*, 2011, **12**, 701.
72. D. R. Kryscio and N. A. Peppas, *Acta Biomater.*, 2012, **8**, 461.
73. C. Gonzalez-Chomon, A. Concheiro and C. Alvarez-Lorenzo, *Ther. Delivery*, 2013, **4**, 1.
74. N. A. Peppas, B. Ekerdt and M. Gomez-Burgaz, Method and process for the production of multi-coated recognition and releasing systems, *U. S. Pat.*, 20090232858, 2009.
75. Q. Zhang, L. Zhang, P. Wang and S. Du, *J. Pharm. Sci.*, 2014, **103**, 643.
76. Y. Ma, Y. Zhang, M. Zhao, X. Guo and H. Zhang, *Chem. Commun.*, 2012, **48**, 6217.
77. L. Xu, J. Pan, J. Dai, X. Li, H. Hang, Z. Cao and Y. Yan, *J. Hazard. Mater.*, 2012, **233–234**, 48.
78. J. Pan, H. Hang, X. Dai, J. Dai, P. Huo and Y. Yan, *J. Mater. Chem.*, 2012, **22**, 17167.
79. S. F. Xu, J. H. Li, X. L. Song, J. Liu, H. Z. Lu and L. X. Chen, *Anal. Methods*, 2013, **5**, 124.
80. L. Fang, S. Chen, X. Guo, Y. Zhang and H. Zhang, *Langmuir*, 2012, **28**, 9767.
81. R. Cheng, F. Feng, F. Meng, C. Deng, J. Feijen and Z. Zhong, *J. Controlled Release*, 2011, **152**, 2.
82. F. Q. Schafer and G. R. Buettner, *Free Radical Biol. Med.*, 2001, **30**, 1191.
83. P. Kuppusamy, H. Li, G. Ilangovan, A. J. Cardounel, J. L. Zweier, K. Yamada, M. C. Krishna and J. B. Mitchell, *Cancer Res.*, 2002, **62**, 307.
84. J. F. Quinn, M. R. Whittaker and T. P. Davis, *Polym. Chem.*, 2017, **8**, 97.
85. J. M. Lambert and C. Q. Morris, *Adv. Ther.*, 2017, **34**, 1015.
86. M. D. White, C. M. Bosio, B. N. Duplantis and F. E. Nano, *Cell. Mol. Life Sci.*, 2011, **68**, 3019.
87. M. T. Calejo, S. A. Sande and B. Nystrom, *Expert Opin. Drug Delivery*, 2013, **10**, 1669.
88. http://celsion.com/thermodox/; accessed April 2018.
89. D. Needham, J. Y. Park, A. M. Wright and J. Tong, *Faraday Discuss.*, 2013, **161**, 515.
90. T. J. Mason, *Ultrason. Sonochem.*, 2011, **18**, SI 847.
91. H. Zhang, H. Xia, J. Wang and Y. Li, *J. Controlled Release*, 2009, **139**, 31.
92. G. A. Husseini, G. D. Myrup, W. G. Pitt, D. A. Christensen and N. Y. Rapoport, *J. Controlled Release*, 2000, **69**, 43.
93. N. Rapoport, in *Smart Nanoparticles in Nanomedicine*, ed. R. Arshady and K. Kono, Kentus Books, London, 2006, pp. 305–362.

94. B. J. Staples, W. G. Pitt, B. L. Roeder, G. A. Husseini, D. Rajeev and G. B. Schaalje, *J. Pharm. Sci.*, 2010, **99**, 3122.
95. P. Kamev and N. Rapoport, *Am. J. Phys.*, 2006, **829**, 543.
96. Y. Zhou, X. X. Han, X. X. Jing and Y. Chen, *Adv. Healthcare Mater.*, 2017, **6**, 1700646.
97. Y. Yang, J. Mu and B. Xing, *WIREs Nanomed. Nanobiotechnol.*, 2017, **9**, e1408.
98. C. Alvarez-Lorenzo, L. Bromberg and A. Concheiro, *Photochem. Photobiol.*, 2009, **85**, 848.
99. Z. Cao, Q. Bian, Y. Chen, F. Liang and G. Wang, *ACS Macro Lett.*, 2017, **6**, 1124.
100. M. Mathiyazhakan, C. Wiraja and C. Xu, *Nano-Micro Lett.*, 2018, **10**, 10.
101. L. Vigderman and E. R. Zubarev, *Adv. Drug Delivery Rev.*, 2013, **65**, 663.
102. J. Yang, J. Lee, J. Kang, S. J. Oh, H. J. Ko, J. H. Son, K. Lee, J. S. Suh, Y. M. Huh and S. Haam, *Adv. Mater.*, 2009, **21**, 4339.
103. C. Kojima, H. Kawabata, A. Harada, H. Hirunaka and K. Kono, *Chem. Lett.*, 2013, **42**, 612.
104. S. Cabana, C. S. Lecona-Vargas, H. Melendez-Ortiz, A. Contreras-Garcia, S. Barbosa, P. Taboada, B. Magarinos, E. Bucio, A. Concheiro and C. Alvarez-Lorenzo, *J. Drug Delivery Sci. Technol.*, 2017, **42**, 245.
105. R. Müller, H. Steinmetz, R. Hiergeist and W. Gawalek, *J. Magn. Magn. Mater.*, 2004, **276**, 272.
106. M. Arruebo, R. Fernández-Pacheco, M. R. Ibarra and J. Santamaría, *Nano Today*, 2007, **2**, 22.
107. C. S. S. R. Kumar and F. Mohammad, *Adv. Drug Delivery Rev.*, 2011, **63**, 789.
108. M. L. Etheridge, S. A. Campbell, A. G. Erdman, C. L. Haynes, S. M. Wolf and J. McCullough, *Nanomedicine: NMB*, 2013, **9**, 1.
109. E. Alphandery, P. Grand-Dewyse, R. Lefevre, C. Mandawala and M. Durand-Dubief, *Expert Rev. Anticancer Ther.*, 2015, **15**, 1233.
110. E. Ruiz-Hernandez, A. Baeza and M. Vallet-Regi, *ACS Nano*, 2011, **5**, 1259.
111. Y. Chen, H. Chen and J. Shi, *Adv. Mater.*, 2013, **25**, 3144.
112. S. Murdan, *J. Controlled Release*, 2003, **92**, 1.
113. S. Kagatani, T. Shinoda, Y. Konno, M. Fukui, T. Ohmura and Y. Osada, *J. Pharm. Sci.*, 1997, **86**, 1273.
114. S. A. Agnihotri, R. V. Kulkarni, N. N. Mallikarjuna, P. V. Kulkarni and T. M. Aminabhavi, *J. Appl. Polym. Sci.*, 2005, **96**, 301.
115. N. Paradee, A. Sirivat, S. Niamlang and W. Prissanaroon-Ouajai, *J. Mater. Sci. Mater. Med.*, 2012, **23**, 999.
116. Y. Zhao, A. C. Tavares and M. A. Gauthier, *J. Mat. Chem. B*, 2016, **4**, 3019.
117. R. Okner, M. Oron, N. Tal, D. Mandler and A. J. Domb, *Mater. Sci. Eng., C*, 2007, **27**, 510.
118. Y. Cho, R. Shi, A. Ivanisevic and R. B. Borgens, *Nanotechnology*, 2009, **20**, 275102.
119. B. C. Thompson, R. T. Richardson, S. E. Moulton, A. J. Evans, S. O'Leary, G. M. Clark and G. G. Wallace, *J. Controlled Release*, 2010, **141**, 161.
120. D. Ge, X. Tian, R. Qi, S. Huang, J. Mu, S. Hong, S. Ye, X. Zhang, D. Li and W. Shi, *Electrochim. Acta*, 2009, **55**, 271.
121. S. Mongkolkitikul, N. Paradee and A. Sirivat, *Eur. J. Pharm. Sci.*, 2018, **112**, 20–27.
122. S. Pairatwachapun, N. Paradee and A. Sirivat, *Carbohydr. Polym.*, 2016, **137**, 214.

16 Review of Smart Mechanochromic and Metamaterials

Mohsen Shahinpoor

Mechanical Engineering Dept., University of Maine, USA
Email: shah@maine.edu

16.1 Introduction

This chapter reviews two recent families of smart materials, namely mechanochromic materials and mechanical metamaterials. Mechanochromic materials change their optical properties, in particular their photoluminescence characteristics, if subjected to mechanical loading or through interactions with their environment. Chemical and physical molecular changes across various length scales and the rearrangement of molecular chemical bonds to modifications in molecular arrangements in the nanometers regime generally trigger mechanochromic characteristics. Metamaterials are defined as materials that are not ordinarily produced in nature. Note that "meta" means "beyond", and metamaterials have properties that go beyond conventional materials. Metamaterials are nanocomposite materials made up of a periodically repeating micro or nano units of metals, alloys and plastics that exhibit properties different from the natural properties of the participating materials. In the following sections, these families of mechanochromic and metamaterials are further described and elaborated on.

Fundamentals of Smart Materials
Edited by Mohsen Shahinpoor
© The Royal Society of Chemistry 2020
Published by the Royal Society of Chemistry, www.rsc.org

16.2   Introduction to Mechanochromic Materials

Mechanochromic materials are capable of changing their optical properties, such as changing their color in response to stress and strain. The observation of mechanical stresses in materials according to their color changes has many applications in structural health monitoring, and the biological and medical fields, as well as keeping track of forces and stresses applied to various systems. Polymers or polymer systems have also emerged with unique mechanochromic characteristics that can be used in many applications. These mechanochromic characteristics originate as a result of the rearrangement of chemical and molecular bonds to changes in molecular domains that are approximately hundreds of nanometers in size. There is a lot of interest in these materials for a variety of biological, biomedical, industrial, medicinal and technological applications. Such mechanochromic effects and characteristics will further improve the understanding of strain/stress distributions in solids.

Phenomena such as mechanochromic luminescence, which some materials exhibit by rubbing or through other mechanical loading, such as pressing and shearing, are already well-known.[1] There has also been a lot of interest in mechanochromic materials and their potential for biological and healthcare applications.[2,3] Mechanochromic polymers also present great potential for various applications.[4] Mechanochromic polymers are essentially macromolecular materials that upon being stressed or strained change their color.[4] The visualization of mechanical stresses in polymeric materials is quite useful for use in several engineering and medical applications. For methodologies relating to the 3D printing of mechanochromic materials, see ref. 5. For the design of polymeric materials, which show mechanical stresses *via* a color change due to mechanically induced changes in their intrinsic absorption, emission properties, or changes in molecular structure, see ref. 6.

16.3   Some Examples of Mechanochromic Polymers

Some typical mechanochromic polymers are depicted in Figure 16.1.[7,8] The emission characteristics of a polymer, in particular, photoluminescent oligo (phenylene vinylene)s, may strongly depend on the extent of aggregation of the luminescent guest molecules. This effect may be exploited for the design of molecular sensors integrated into a

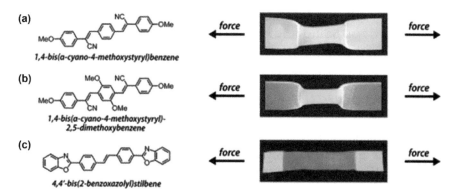

Figure 16.1 Selected images of the corresponding polymer films after tensile deformation recorded under UV light of (a) 1, 4-bis(α-cyano-4-methoxystyryl)benzene, (b) 1,4-bis(α-cyano-4-methoxystyryl)-2,5-dimethoxybenzene in linear low-density poly(ethylene), and (c) 4,4'-Bis(2-benzoxazolyl)stilbene in poly(propylene). (a) and (b) Reproduced from ref. 7 with permission from the American Chemical Society, Copyright 2003. (c) Reproduced from ref. 8 with permission from John Wiley and Sons, © 2005 WILEY-VCH Verlag GmbH & Co. KGaA, Weinheim.

polymer of interest to monitor mechanical deformation. Poly(propylene) (PP) films containing different concentrations of bis(benzoxazolyl) stilbene (BBS) have been prepared by melt processing.

Mechanochromic polymers have been used as strain sensors.[9] In this respect, the authors considered thermoplastic polyurethane, which was mixed with 0.5 wt.% of bis(benzoxazolyl)stilbene dye. The material mechanochromic behavior was exhibited in the form of fluorescence emission when stretched. They[9] correlated the fluorescence response with the mechanical strain for sensing applications. Cellini *et al.*[10] also presented methodologies for strain measurements using mechanochromic polymers.

16.4 Mechanochromic Devices Based on Marine Biological Systems

If disturbed or attacked, jellyfish are capable of changing their appearance underwater by going from a transparent state to an opaque state. Cephalopods are capable of changing their skin color pattern for camouflage as well as communications with other Cephalopods. These methods can be employed towards designing anti-glare screens and color-changing clothing. Squid and octopi use a unique skill to hide from predators on the ocean floor by employing their muscles to expand or expose pigment sacks in their skin to achieve a specific color or pattern. Squid and octopi are also capable of changing their skin texture at will, going from smooth to wrinkled or rippled skin.

16.5 Introduction to Mechanical Metamaterials

Any new property arises as a result of how the repeated units are structurally and periodically arranged according to one another. By this definition, most engineering materials may be used to make metamaterials. Similarly, most composite materials are technically metamaterials, as they contain different materials in a configuration that does not commonly occur in nature. Some of these meta properties include a negative Poisson's ratio, negative elastic modulus, negative refractive index, unusually low thermal conductivity, and high specific energy absorption, in addition to other properties. Mechanical metamaterials can be designed and assembled to have a negative dynamic modulus, photonic bandgaps, superior thermo-electric properties, and high specific energy absorption. In this chapter, micro-/nanostructured materials as mechanical metamaterials are described. The amazing properties of metamaterials are dependent on the design/geometry of the micro/nanoscale structure of the material, which is periodically arranged. Metamaterials were originally developed to manipulate electromagnetic waves and their applications, including super-lenses, slow light, data storage, and optical switching. However, this chapter will mostly focus on mechanical metamaterials rather than electromagnetic metamaterials. For some good references on metamaterials see ref. 11–14.

16.6 Background to Metamaterials

As emphasized in the previous sections, the special properties of metamaterials normally arise as a result of their unique repeated periodic mini, micro or nanostructure, rather than the basic materials they are made from. The earliest study of metamaterials was by Jagadish Chandra Bose, who started researching chiral materials in 1898[11,12] and discussed how chiral molecules consist of all the same components and have "left-handed" and "right-handed" versions.

Additive manufacturing (3D printing) has played a big role in promoting metamaterial technologies. Layer-by-layer stacking of multiple 2D structures can lead to the creation of micro and nanoscale metamaterials (Jae-Hwang Lee[13]). Therefore, additive assembly, manufacturing, and stacking may be preferable, as it is simpler and less costly. Metamaterial manufacturing techniques include angled exposure techniques, interference lithography, direct wire lithography, self-assembly, and 3D printing. An interesting class of metamaterials is pentamode materials, which have a high modulus in 5 of the 6 principal directions, resulting in a high bulk modulus, low shear modulus, and a Poisson's ratio of 0.5 (incompressibility, rubber-like materials). These materials are generally composed of repeated periodic microstructure in the form of conical beams in a diamond-type lattice structure, as shown in Figure 16.2.[14]

Another type of penta-mode material is a gel that can deform in any direction, but resists changes in volume under hydrostatic pressure (Amir Zadpoor[14]). Auxetic meta-materials are also interesting, and they exhibit properties opposite to those of penta-mode metamaterials as they have a small bulk modulus, high shear modulus, and negative Poisson's ratio approaching −1 (Figure 16.3).

Therefore, auxetic metamaterials exhibit lateral expansion under axial tension and lateral contraction under axial compression. Another property auxetic materials exhibit is synclastic curvature, a surface that bends in the same direction in both perpendicular planes. This property allows for auxetic materials to conform to a dome shape without needing seams or joining. Auxetic materials include porous polymeric and metallic foams, microporous polymers, honeycomb, and chiral structures. The applications of auxetic materials include their use as membrane filters with variable

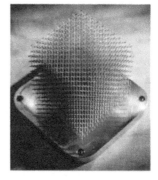

Figure 16.2 Microscale (left) and macroscale (right) depictions of pentamode materials.
Reproduced from ref. 14 with permission from the Royal Society of Chemistry.

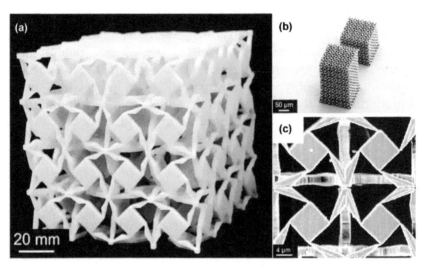

Figure 16.3 Auxetic materials possessing a negative Poisson's ratio on a (a) macroscopic and (b, c) microscopic scale.
Reproduced from ref. 14 with permission from the Royal Society of Chemistry.

permeability, fasteners, shape memory materials, running shoes, bioprosthetics, and acoustic dampeners. The actual effects of auxetic metamaterials typically only have been exhibited in strains that are less than ten percent. The design of 3D auxetic materials that maintain a negative Poisson's ratio under larger strains remains a challenge.

By 2007, experiments that involved a negative refractive index had been conducted by many groups.[14] At microwave frequencies, the first, imperfect invisibility cloak was realized in 2006.

There are various kinds of metamaterials known to us presently. Some of them are as follows:

16.7   Electromagnetic Metamaterials

An electromagnetic metamaterial affects electromagnetic waves that interact with its structural features, which are smaller than the wavelength.

16.8   Elastic Metamaterials

These metamaterials use different parameters to achieve a negative index of refraction in materials that are not electromagnetic.

16.9   Acoustic Metamaterials

Acoustic metamaterials control, direct and manipulate sound in the form of sonic, infrasonic or ultrasonic waves in gases, liquids, and solids.

16.10   Structural Metamaterials

Structural metamaterials have properties that include being crushable and lightweight. Using micro-stereo projection lithography, micro lattices can be created using forms much like trusses and girders.

16.11   Nonlinear Metamaterials

May be fabricated that include some form of nonlinear media, the properties of which change with the power of the incident wave.

16.12   Cloaking Devices

The negative refractive index of a metamaterial is utilized to bend the light around the object in question to make it invisible to the human eye. Therefore, a person looking directly at the object is seeing the projection behind that hidden object.

16.13   Seismic Protection

Seismic metamaterials counteract the adverse effects of seismic waves on human-made structures.

16.14   Antennas

Metamaterial antennas are a class of antennas that use metamaterials to improve performance.

16.15   Absorber

A metamaterial absorber manipulates the loss components of a metamaterials' permittivity and magnetic permeability, to absorb large amounts of electromagnetic radiation.

16.16   Super Lens

A superlens is a two or three-dimensional device that uses metamaterials, usually with negative refraction properties, to achieve resolution beyond the diffraction limit (ideally, infinite resolution).

16.17   Optical Metamaterials

Optical metamaterials are a class of composite materials, the microstructures of which give rise to unusual optical properties, including negative refraction (where the light refracts in the opposite direction to that normally expected).

16.18   Conclusions

The chapter briefly described two novel families of smart materials, **mechanochromic materials** and **mechanical metamaterials**. The chapter further briefly discussed the mechanisms of changing color or other optical characteristics of mechanochromic materials under mechanical stress and strain. The approaches toward such materials have evolved from the use of conjugated polymers that undergo conformational changes of their backbone on the application of mechanical stimuli to the stress-induced dispersion of chromophore aggregates in polymer matrices. The field of mechanochromic materials has seen significant progress, and many opportunities remain. The recent developments in mechanochemistry will lead to a surge in the development of polymeric materials that respond to mechanical deformation in a useful manner.

Metamaterials, on the other hand, are a rich field of emerging materials technology, with applications varying from structures to actuation to optics, and even to invisibility. By this definition, most engineering materials may be used to make metamaterials. These materials will undoubtedly shape the future of technology and engineering. Similarly, most composite materials are technically metamaterials, as they contain different materials in a configuration that does not commonly occur in nature. Some of these meta properties include a negative Poisson's ratio, negative elastic modulus, negative refractive index, unusually low thermal conductivity, and high specific energy absorption, in addition to other properties. Mechanical metamaterials have shown extraordinary properties, such as negative dynamic modulus, photonic bandgaps, superior thermoelectric properties, and high specific energy absorption.

In this chapter, micro-/nanostructured mechanical metamaterials are described. The amazing properties of metamaterials are dependent on the design/geometry of the micro/nanoscale structure of the material, which is periodically arranged.

Homework Problems

Homework Problem 16.1

What are the basic characteristics of mechanochromic materials?

Homework Problem 16.2

What is the basic microstructural arrangement in mechanochromic materials?

Homework Problem 16.3

Where do mechanochromic materials originate from in terms of chemical and physical molecular changes across various length scales, such as the rearrangement of molecular chemical bonds to modifications in molecular arrangements in the nanometer regime?

Homework Problem 16.4

What is the fundamental difference between metamaterials and natural materials?

Homework Problem 16.5

What are the basic characteristics of metamaterials?

Homework Problem 16.6

What is the basic microstructural arrangement in metamaterials?

Homework Problem 16.7

What are the basic characteristics of image cloaking pentamode materials?

Homework Problem 16.8

What are the basic characteristics of auxetic cloaking materials?

References

1. X. Sun, X. Zhang, X. Li, S. Liu and G. Zhang, A mechanistic investigation of mechanochromic luminescent organoboron materials, *J. Mater. Chem.*, 2012, **22**(33), 17332–17339.
2. Y. Jiang, An outlook review: Mechanochromic materials and their potential for biological and healthcare applications, *Mater. Sci. Eng., C*, 2014, **45**, 682–689.
3. J.-H. Lee, P. S. Jonathan and L. T. Edwin, Micro-/Nanostructured Mechanical Metamaterials, *Adv. Mater.*, 2012, **24**(36), 4782–4810.
4. C. Weder, Mechanochromic Polymers, in *Encyclopedia of Polymeric Nanomaterials*, Springer, Berlin/Heidelberg, 2014, pp. 1–11.
5. G. I. Peterson, M. B. Larsen, M. A. Ganter, D. W. Storti and A. J. Boydston, 3D-Printed Mechanochromic Materials, *ACS Appl. Mater. Interfaces*, 2015, **7**(1), 577–583.
6. L. Céline Calvino, C. Neumann and S. S. Weder, Approaches to polymeric mechanochromic materials, *Inc. J. Polym. Sci., Part A: Polym. Chem.*, 2017, **55**(4), 640–652.
7. B. R. Crenshaw and C. Weder, Deformation-Induced Color Changes in Melt-Processed Photoluminescent Polymer Blends, *Chem. Mater.*, 2003, **15**(25), 4717–4724.
8. A. Pucci, M. Bertoldo and S. Bronco, Luminescent bis(benzoxazolyl)stilbene as a molecular probe for poly(propylene) film deformation, *Macromol. Rapid Commun.*, 2005, **26**(13), 1043–1048.
9. F. Cellini, S. Khapli, S. D. Peterson and M. Porfiri, Mechanochromic polyurethane strain sensor, *Appl. Phys. Lett.*, 2014, **105**, 061907.
10. F. Cellini, S. J. Osma, S. D. Peterson and M. Porfiri, A Mechatronics-Based Platform for In Situ Strain Measurement Through Mechanochromic Polymers, *IEEE/ASME Trans. Mech.*, 2016, **21**(6), 2989–2995.
11. Wikipedia Contributors, Metamaterial, Wikipedia, The Free Encyclopedia. Wikipedia, The Free Encyclopedia, 5 Apr. 2017. Web. 2 May, 2017.
12. Wikipedia Contributors, Chirality, Wikipedia, The Free Encyclopedia, 7 Apr. 2017. Web. 2 May, 2017.
13. J.-H. Lee, P. S. Jonathan and L. T. Edwin, Micro-/Nanostructured Mechanical Metamaterials, *Adv. Mater.*, 2012, **24**(36), 4782–4810.
14. A. A. Zadpoor, Mechanical Meta-materials, *Mater. Horiz.*, 2016, **3**(5), 371–381.

17 Review of Ionic Polymer–Metal Composites (IPMCs) as Smart Materials

Mohsen Shahinpoor

Mechanical Engineering Dept., University of Maine, USA
Email: shah@maine.edu

17.1 Introduction

This chapter presents a review on ionic polymer–metal composites (IPMCs) with conjugated cations that can be redistributed if they are placed in an electric field and consequently act as distributed nanoactuators. Without an applied field, they can also act as distributed self-powered nanosensors and energy harvesters. This review also briefly presents the manufacturing methodologies and fundamental properties and characteristics of IPMCs.

This chapter also presents a phenomenological model of IPMCs based on linear irreversible thermodynamics. In this formulation, the underlying actuation and sensing mechanisms are described by two driving forces, an electric field, E, a solvent pressure gradient, Δp, and two fluxes, the electric current density, J, and the ionic + plasticizer flux, Q.

Polymers containing equilibrated and conjugated ions within their molecular networks present a great opportunity to create smart nanocomposites with distributed nanoactuation, nanosensing, nano transduction and energy harvesting capabilities for a variety of industrial, scientific and medical applications. For a comprehensive introduction

Fundamentals of Smart Materials
Edited by Mohsen Shahinpoor
© The Royal Society of Chemistry 2020
Published by the Royal Society of Chemistry, www.rsc.org

to these types of smart and intelligent materials, the reader is referred to work by Shahinpoor and Schneider[1] and Shahinpoor.[2,3] The reader is further referred to studies by Shahinpoor, Kim, and Mojarrad[4–6] and review articles by Shahinpoor and Kim.[5–8] Gel-Based,[1,9–12] and chitosan-based[11] conductor composites have also been considered as electrically active composite smart materials.

IPMCs are entirely ionic capacitive actuators and sensors. With increasing ion content, the number density of the IPMC increases, leading to the overlapping of the restricted mobility regions and the formation of clusters. Such clusters, for example, in perfluorinated ionic polymers such as Nafion[®], are 3 to 5 nm in size.[2,3] Ionic aggregates, in a polymer matrix of low polarity, are formed through the association of ion pairs in an ionomer. It must be noted that widespread electro-chemical processes and devices utilize poly(perfluorosulfonic acid) ionic polymers. These materials exhibit[1,2,13,14] good chemical stability, remarkable mechanical strength, good thermal stability, and high electrical conductivity when sufficiently hydrated and made into a nanocomposite with a conductive phase such as metals, conductive polymers or graphite.

As described in ref. 13–17 a number of physical models have been developed to understand the mechanisms of water and ion transport in ionic polymers and membranes. It has also been well established[2] that anions are tethered to the polymer backbone and cations (H^+, Na^+, Li^+) are mobile and solvated by polar or ionic liquids within nanoclusters of 3–5 nm in size.

17.2   Three-dimensional Fabrication of IPMCs

As reported in studies by Shahinpoor, Kim and Mojarrad,[4] for ion-containing polymers in a nanocomposite, a conductor can be manufactured three-dimensionally to any complex shape, as for example shown in Figure 17.1.

There are essentially two ways to manufacture three-dimensional ionic polymeric nanocomposites with a conductor phase. One is to use a liquid form of the polyelectrolyte in an alcohol, such as liquid Nafion[®] in isopropyl alcohol. By meticulously evaporating the solvent (isopropyl alcohol) out of the solution, recast ionic polymer can be obtained for any desired shape and dimensions and, particularly, thickness.

The other method is to use a precursor resin (XR Resin, DuPont) and hydrolyze it in KOH and then make it into a composite with a conductive

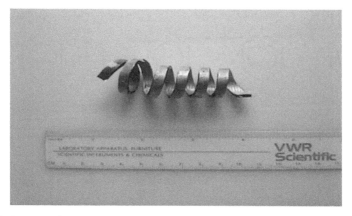

Figure 17.1   A coil-type ionic polymeric nanocomposite with gold electrodes for linear nanosensing and nanoactuation. Reproduced from ref. 1 with permission from the Royal Society of Chemistry.

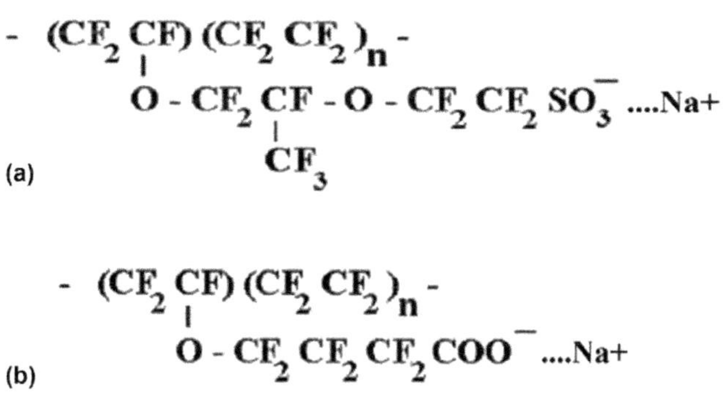

$$- \, (CF_2 \, CF) \, (CF_2 \, CF_2)_n -$$
$$O - CF_2 \, CF - O - CF_2 \, CF_2 \, SO_3^- \, ...Na+$$
$$CF_3$$

(a)

$$- \, (CF_2 \, CF) \, (CF_2 \, CF_2)_n -$$
$$O - CF_2 \, CF_2 \, CF_2 \, COO^- \, ...Na+$$

(b)

Figure 17.2   Perfluorinated acid polymers used in ion exchange and REDOX processes to convert to IPMCs (a, Nafion®, DuPont) and (b, Flemion®, Asahi).

phase. Manufacturing of ionic polymer nanocomposites begins with the selection of an appropriate ionic polymeric material. Often, ionic polymeric materials are manufactured from polymers that consist of fixed covalent, ionic groups. For example, perfluorinated polyalkenes with short side-chains terminated by ionic groups [typically sulfonate or carboxylate ($SO_3^- \, ... \, Na^+$ or $COO^- \, ... \, Na^+$) for cation exchange or ammonium cations for anion exchange (see Figures 17.2a and b)] can be manufactured using standard techniques.[4]

The large polymer backbones determine their mechanical strength. Short side-chains provide ionic groups that interact with water and the passage of appropriate ions. Additionally, there are styrene/divinylbenzene-based polymers, in which the ionic groups have been

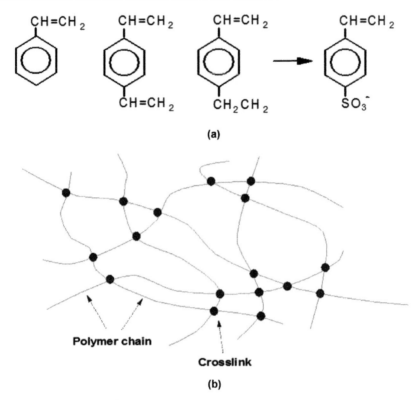

(a)

(b)

Figure 17.3 Styrene/divinylbenzene-based ion exchange materials (a) and their molecular structural form (b).
Reproduced from ref. 1 with permission from the Royal Society of Chemistry.

substituted from the phenyl rings where the nitrogen atom is fixed to an ionic group (Figure 17.3).

Exercise No. 17.1 What is the fundamental molecular difference between sulfonic and carboxylic ion exchange polymeric membrane?

Answer to Exercise 17.1 The main difference is that the ionic pendant groups are $SO^{3-}\ldots Na^+$, where Na^+ can be any metallic cation or hydrogen cation, while in carboxylic membranes the ionic pendant branches are in the form of $COO^-\ldots Na^+$, as shown in Figure 17.2.

The essential steps in the manufacturing of ionic polymeric nanocomposites are: roughening the material surface to increase the surface density during ion exchange and then placing it in a water-soluble metallic salt solution to oxidize it and then in a reducing solution to reduce it in a typical REDOX operation.

The second step is to incorporate the ion exchanging process using a water-soluble metal complex solution such as tetra-amine platinum chloride hydrate as an aqueous platinum, such as ($[Pt(NH_3)_4]Cl_2$ and $[Pt(NH_3)_6]Cl_4$), or gold complexes such as dichlorophenanthroline gold(III) chloride, $[Au(phen)Cl_2]Cl$ or ammonium tetrachloroaurate(III) hydrate $NH_4AuCl_4 \cdot xH_2O$ in solution. Although the equilibrium conditions depends on the types of charge on the metal complex, such complexes have found to make good electrodes.

The third step in manufacturing ionic polymer platinum composites is to place the oxidized IPMC in a reducing solution such as sodium or lithium borohydride at a favorable temperature (*i.e.*, 60 °C) for a couple of hours. Figure 17.4a, b, c, and d depict an ionic polymer near the boundary distribution of electrodes, resembling fractal configurations of nanoparticles of the reduced metal within the polymer macromolecular network.

It has been established[2] that surface conductivity of ionic polymer conductor nanocomposites plays an essential role in the ultimate performance of the distributed nanosensor and nanoactuator.

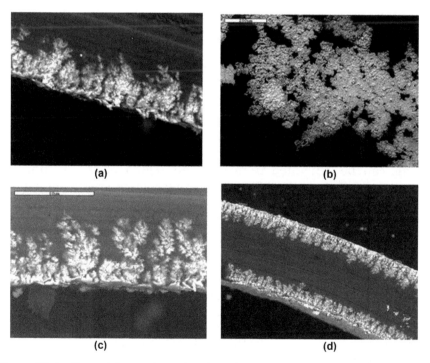

(a)   (b)

(c)   (d)

Figure 17.4   Fractal formation of reduced nanoparticles of metal within the polymeric macromolecular network, (a) near the boundary formation of nanoparticles, (b) close-up picture of fractal formation, (c) penetration of fractal nanoparticle formation into the polymeric network, and (d) an SEM image of a 200 micron thick ionic polymeric membrane metal-plated on both sides.

Atomic force microscopy (AFM) was used for the surface characterization of ionic polymer platinum nanocomposites. Its capability to directly image the surface of the polymeric composite provided detailed information, with a resolution of a few nanometers.

The surface is characterized by the granular appearance of platinum metal, with a peak/valley depth of approximately 50 nm. Figure 17.5 depicts an AFM picture of the surface morphology of the polymeric nanocomposite.

Exercise No. 17.2 What is the reason behind surface roughening with sandpaper or glass beads of ion exchange membranes during the manufacturing of IPMCs?

Answer to Exercise 17.2 The reason is to increase the surface density (actual available surface area divided by the projection of that two-dimensionally) to further enhance and accelerate the diffusion of metallic salt during oxidation and diffusion of reducing agent during the reduction of IPMCs.

Figure 17.6 shows a TEM image on the penetrating edge of the ionic polymer nanocomposite. A small piece of sample was carefully prepared and ion-beam treated. The average particle size was found to be around 47 nm.

Exercise No. 17.3 What is the reason behind claiming that IPMCs are distributed nanoactuators, nanosensors, and nano energy harvesters?

Answer to Exercise 17.3 The reason is that the chemically plated electrodes do provide distributed electric field to everywhere on the ionic membranes. The distributed electrodes mean that sensing signals can be generated from any nanoregion of the membranes because local electrodes will transmit the local nano signal, and it can be recorded.

Figure 17.7 further shows the distribution of nanoparticles in the near boundary region of the ionic polymeric nanocomposite. Once these manufacturing steps are concluded, the outcome is a nanocomposite with surface electrodes because the reduction of metals occurs within the ionic nanoclusters (Figure 17.8), as described in ref. 2–4.

Exercise No. 17.4 How can one prevent metallic nanoparticles from being reduced on the molecular network to coalesce and become microparticles?

Answer to Exercise 17.4 One can use soap-like dispersing materials such as detergents and soaps to prevent the nanoparticles from coalescing. One powerful soap-like dispersant is polyvinylpyrrolidone or (PVP).

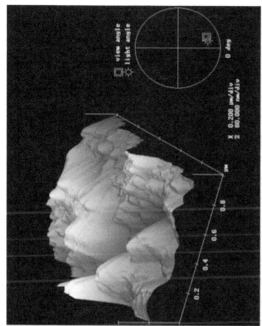

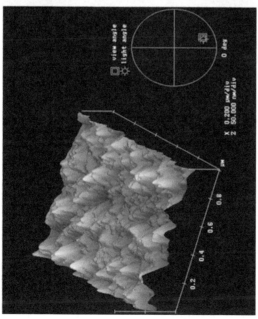

Figure 17.5 Atomic force microscopy images taken of the surface electrodes of a couple of typical ionic polymer nanocomposite samples. The scanned area is 1 μm^2.

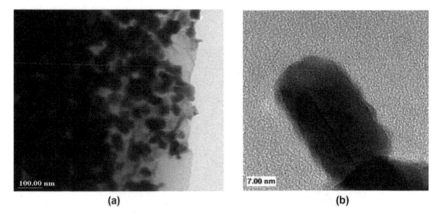

(a)   (b)

Figure 17.6   Two TEM photos (a) and (b) of images of platinum nanoparticles within the IPMC network.

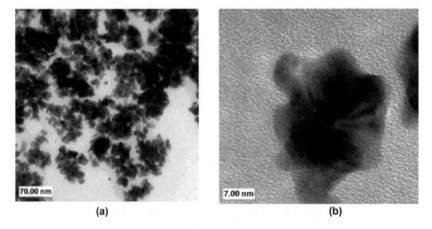

(a)   (b)

Figure 17.7   (a) Distribution of platinum nanoparticles within the ionic polymeric macromolecular network and (b) a typical platinum nanoparticle.

The polymeric nanocomposite is now ready to be tested for nanosensing and nanoactuation, as described in the following sections.

Exercise No. 17.5 Describe the reasons why during the reduction phase of chemical plating of Nafion with a metal, nanoparticles of reduced metals are manufactured.

Answer to Exercise 17.5 There are two reasons why these nanoparticles are produced and get embedded within the molecular structure of the ion-containing membrane. One reason is that the chemical structures of perfluorinated ion exchange polymers possess charged nanoclusters of 3–5 nm in diameter for subsequent metallic reduction and production of nanoparticles. The second reason is the

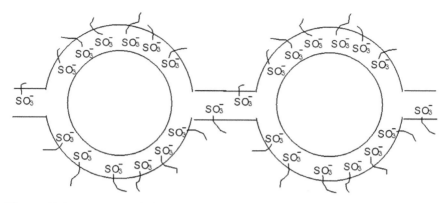

Figure 17.8   Suggested chemical structure of a perfluorinated ion exchange polymer showing the formation of charged nanoclusters of 3–5 nm in diameter for subsequent metallic reduction and production of nanoparticles.

presence of PVP as a dispersant in the reducing solution, preventing the nanoparticles from coalescing and enlarging to microparticles.

17.3   Electrically-induced Robotic Actuation

In perfluorinated sulfonic acid polymers there are relatively few fixed ionic groups, and they can create hydrophilic nano-channels, called *cluster networks* [Gierke and Hsu[16] and Gierke, Munn, and Wilson[17]]. Once an electric field is applied to such a network, the conjugated and hydrated cations rearrange to accommodate the local electric field, and thus the network deforms, as shown in Figure 17.9.

Once an electric field is applied on an ionic polymeric nanocomposite layer either in a cantilever form or flapping-wing configuration by placing the electrodes in the middle of the ionic polymeric nanocomposite strip, the hydrated cations migrate towards the cathode side to accommodate the local electric field. This migration creates tension near the cathode and compression near the anode and thus creates a pressure gradient across the thickness of the strip, thus causing the strip to undergo bending deformation, as shown in Figures 17.10–17.12, under a small applied electric field in the order of 10 s of volts per millimeter.

Figures 17.13 and 17.14 depict typical dynamic force and deflection characteristics of cantilever samples of ionic polymeric nanocomposites.

Exercise No. 17.6 Describe the reasons why one can produce different kinds of deformation in IPMCs such as bending, rolling, twisting, turning, or whirling *via* the application of low voltage actuation.

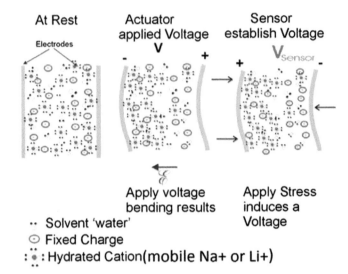

Figure 17.9   Sensing and actuation mechanisms in an ionic polymer nano-composite due to internal ionic redistribution either due to mechanical deformation (sensing and energy harvesting) or an applied electric field.

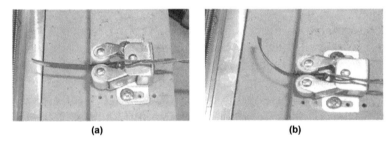

Figure 17.10   Deflection of (a) a strip (1 cm×4 cm×300 microns) of an ionic polymeric nanocomposite (b) under a DC voltage at 2 volts.

Figure 17.11   Typical deformation of strips (1 cm×8 cm×0.34 mm) of ionic polymers under a step voltage of 4 V. Reprinted from ref. 1 with permission from The Royal Society of Chemistry.

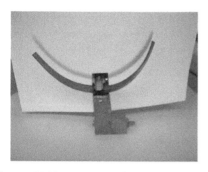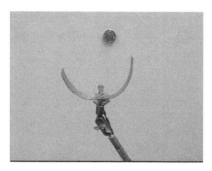

Figure 17.12 Ionic polymer nanocomposites in flapping pair of wings configurations.

Answer to Exercise 17.6 The reason is that depending on the geometry of external activating electrodes printed or placed on various surfaces of an IPMC strip by masking or digital printing, the migration of internal cations within the ionic polymer leads to specialized, localized deformations with specific configurations. Thus, one can literally produce all kinds of motion with IPMCs, such as bending, rolling, twisting, turning or whirling.

Note that a typical set up to measure the force uses a load cell and measures the blocking force of the actuator in a cantilever configuration, as depicted in Figure 17.15.

Exercise No. 17.7 Describe the reasons why one can produce different kinds of self-powered sensing signals from IPMCs by bending, rolling, twisting, turning, stretching, or whirling.

Answer to Exercise 17.7 The reason is that depending on the imposed geometry of deformation, the redistribution of cations can create a specific output voltage signal based on Poisson–Nernst–Planck field equations.

17.4 Distributed Nanosensing and Transduction

IPMCs are excellent self-powered sensors and energy harvesters. The material produces a potential that is dependent on the rate of deformation and the magnitude of deformation. This output can be used for a wide verity of engineering applications. Since the material is soft, it can be placed in fabric and used as a movement sensor embedded in smart fabrics. Shahinpoor and Schneider,[1] Shahinpoor, Kim and Mojarrad,[2] Shahinpoor and Kim,[3] Shahinpoor,[18] Sadeghipour, Salomon and Neogi[23] and deGennes, Shahinpoor, Okumura and Kim[24] have presented the background on the sensing and transduction

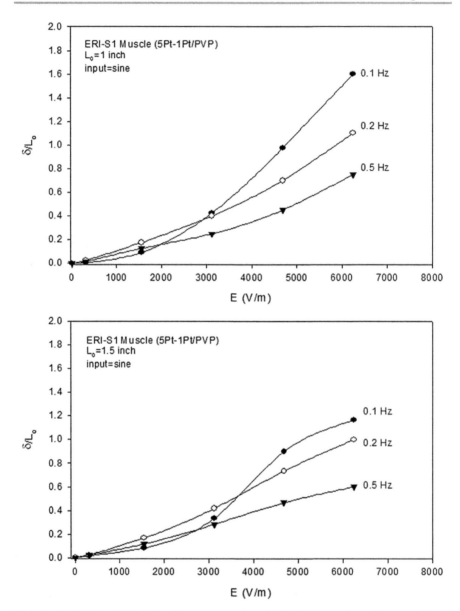

Figure 17.13 Typical deflection *versus* electric field for various frequencies of excitation (δ: tip deflection, L_o: effective cantilever beam length).

properties of ionic polymer conductor composites. Shahinpoor[19,20] and Shahinpoor and Mojarrad[21,22] reported the **ionic flexogelectric** effect, in which the loading of ionic polymer conductor composite strips creates an output voltage like that of a dynamic sensor or a transducer, converting mechanical energy to electrical energy, as shown in Figure 17.16.

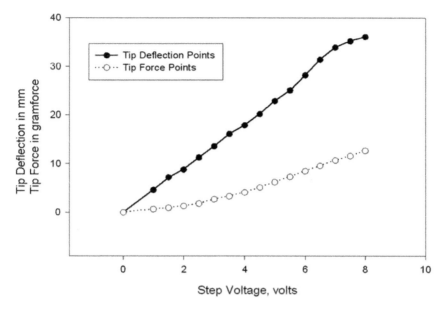

Figure 17.14 Variation of tip blocking force and the associated deflection if allowed to move *versus* the applied step voltage for a 1 cm×4 cm×0.3 mm IPMC sample weighing about 0.25 g in a cantilever configuration generating up to 10 g of blocking force, giving rise to a force density of about 40.

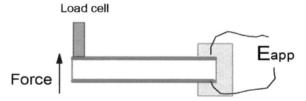

Figure 17.15 Cantilever and load cell configuration for measuring the tip blocking force of IPMC samples.

High-speed and low-speed sensing are depicted in Figures 17.17 and 17.18.

17.5 Modeling and Simulation

Nobel Laureate Pierre de Gennes, Okumura, Shahinpoor and Kim[24] presented the first phenomenological theory for sensing and actuation in IPMCs based on Onsager formulations and linear irreversible thermodynamics Asaka and Oguro[25] have also presented a model for the bending of IPMC strips by electric stimuli and the electro–osmotic drag term in transport equations.

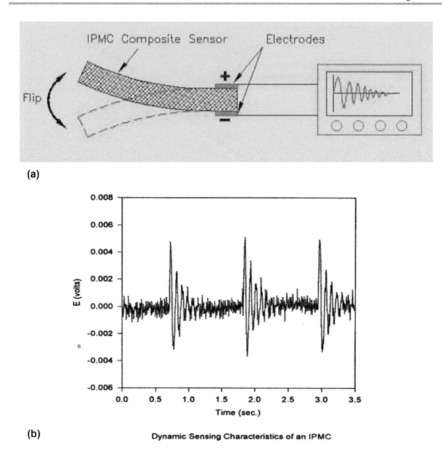

(a)

(b)   **Dynamic Sensing Characteristics of an IPMC**

Figure 17.16   A typical voltage response (b) of an IPMC strip (1 cm×4 cm× 0.2 mm) under oscillatory mechanical excitations in a cantilever configuration (a).

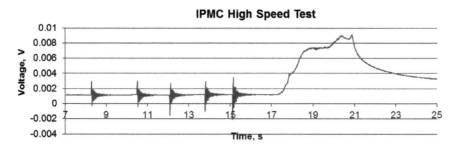

Figure 17.17   High-speed oscillatory and low-speed sensing of IPMCs.

A recent comprehensive review by Shahinpoor and Kim[5] on modeling and simulation of ionic polymeric, artificial muscles discusses the various modeling approaches for the understanding of the mechanisms

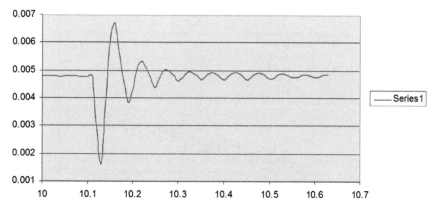

Figure 17.18   High-speed oscillatory sensing of IPMC voltage (mv) *vs.* time (s) for a short time interval.

of sensing and actuation of ionic polymers and the notion of ion mobility. Under quasi-*static conditions*, a simple description of the *mechanoelectric effect* is possible based upon two forms of transport: *ion transport* (with a current density, J, normal to the material) and *solvent transport* (with a flux, Q, *we can assume that this term is water flux*).

The conjugate forces include the electric field, E, and the pressure gradient, $-\nabla p$. The resulting equation has the concise form of:

$$J(x,y,z,t) = \sigma E(x,y,z,t) - L_{12} \nabla p(x,y,z,t) \tag{17.1}$$

$$Q(x,y,zt) = L_{21} E(x,y,z,t) - K \nabla p(x,y,z,t) \tag{17.2}$$

where σ and K are the material electric conductance and the Darcy permeability, respectively. A cross coefficient is usually $L = L_{12} = L_{21}$. The simplicity of the above equations provides a compact view of the underlying principles of actuation, transduction, and sensing of the ionic polymer nanocomposites, also shown in Figures 17.13–17.15 and 17.18.

For the actuation mode (Figures 17.13 and 17.14), the *direct* effect is measured with electrodes that are impermeable to ion species flux, and thus we have $Q = 0$. Thus eqn (17.3) results:

$$\nabla p(x,y,z,t) = \frac{L}{K} E(x,y,z,t) \tag{17.3}$$

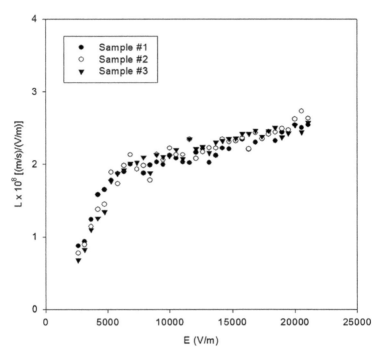

Figure 17.19   Experimental determination of the Onsager Coefficient *L*.

This pressure gradient $\nabla p(x,y,z,t)$ will, in turn, induce a curvature $\underset{\sim}{\kappa}$ proportional to $\nabla p(x,y,z,t)$. The relationships between the curvature $\underset{\sim}{\kappa}$ and pressure gradient $\nabla p(x,y,z,t)$ are described in work by de Gennes, Okumura, Shahinpoor, and Kim.[24] Note that $(1/\rho_c) = \mathbf{M(E)}/\mathrm{YI}$, where $\mathbf{M(E)}$ is the local induced bending moment, which is a function of the imposed electric field **E**. Furthermore, *Y* is the Young's modulus (elastic stiffness) of the strip, which is a function of the hydration, *H*, of the ionic polymer metal nanocomposite, and *I* is the moment of inertia of the strip.

Note that locally, M(**E**) is related to the pressure gradient across the thickness *t** such that if the osmotic pressure is P in tension and −P in compression, then in a simplified scalar format:

$$\nabla p(x,y,z,t) = (2P/t^*) = (\mathbf{M}/I) = Y/\rho_c = Y \underset{\sim}{\kappa}. \qquad (17.4)$$

Now from eqn (17.4), it is clear that the curvature $\underset{\sim}{\kappa}_E$ is related to the imposed electric field **E** by:

$$\underset{\sim}{\kappa}_E = (\mathrm{L/KY}) \underset{\sim}{E} \qquad (17.5)$$

Based on this simplified model, the tip bending deflection δ_{max} of an IPMC strip of length l_g should be almost linearly related to the imposed electric field due to the approximation that:

$$\underset{\sim}{\kappa}_E \cong \left[2\,\underset{\sim}{\delta}_{max} \middle/ \left(l_g^2 + \delta_{max}^2 \right) \right] \cong 2\,\underset{\sim}{\delta}_{max}/l_g^2 \cong (L/KY)\,\underset{\sim}{E} \qquad (17.6)$$

The experimental deformation characteristics depicted in Figures 17.13–17.15 and 17.18 are consistent with the above predictions, which were obtained using the above linear irreversible thermodynamics formulation.

The formulations are also consistent with eqn (17.5) and (17.6) under steady state conditions and have been used to estimate the value of the Onsager coefficient, L, to be in the order of 10^{-8} $m^2 V s^{-1}$, as shown in Figure 17.19.

Here, we have used a low-frequency electric field to minimize the effect of loose water back diffusion under a DC electric field. Other parameters have been experimentally measured to be $K \sim 10^{-18}$ m^2 CP, $\sigma \sim 1$ A mV^{-1}, or S m^{-1}.

This description concludes our coverage of ionic polymer conductor nanocomposites as multi-functional intelligent materials with distributed nanosensing, nano actuation, and nano transduction capabilities.

17.6   Conclusions

A review of IMPCs as smart multi-functional materials capable of actuation, energy harvesting and sensing was introduced in this chapter, which also briefly presented the manufacturing methodologies and fundamental properties and characteristics of these IPMCs.

A phenomenological model (Onsager model) of the underlying actuation and sensing mechanisms was also presented based on linear irreversible thermodynamics with two driving forces, an electric field **E** and a solvent pressure gradient **Δp** and two fluxes, the electric current density **J** and the ionic + plasticizer flux **Q**. See also references 21 and 25–29 in connection with micro-grippers and sensing patents.

Homework Problems

Homework Problem 17.1

Describe the bending mechanism of an IPMC under an electric field.

Homework Problem 17.2

How can you create a parallel jaw robotic gripper with a strips of IPMCs?

Homework Problem 17.3

Describe the mechanism of sensing and energy harvesting of an IPMC.

Homework Problem 17.4

Describe a procedure to measure the force density of IPMC cantilever strips.

References

1. *Intelligent Materials*, ed. M. Shahinpoor and H.-J. Schneider, Royal Society of Chemistry, Cambridge, 2008.
2. M. Shahinpoor, *Ionic Polymer–Metal Composites (IPMCs): Smart Multi-Functional Materials and Artificial Muscles*, Royal Society of Chemistry, Cambridge, vol. II, 2016.
3. M. Shahinpoor, *Ionic Polymer–Metal Composites (IPMCs): Smart Multi-Functional Materials and Artificial Muscles*, Royal Society of Chemistry, Cambridge, vol. I, 2016.
4. M. Shahinpoor, K. J. Kim and M. Mojarrad, *Artificial Muscles: Applications of Advanced Polymeric Nano-Composites*, CRC Press, Taylor & Francis Publishers, London, 2007.
5. M. Shahinpoor and K. J. Kim, Ionic Polymer-Metal Composites – I. Fundamentals, *Smart Mater. Struct. Int. J.*, 2001, **10**, 819–833.
6. K. J. Kim and M. Shahinpoor, Ionic Polymer-Metal Composites – II. Manufacturing Techniques, *Smart Mater. Struct. (SMS), Inst. Phys. Publ.*, 2003, **12**(1), 65–79.
7. M. Shahinpoor and K. J. Kim, Ionic Polymer-Metal Composites – III. Modeling and Simulation As Biomimetic Sensors, Actuators, Transducers, and Artificial Muscles, *Smart Mater. Struct. Int. J.*, 2004, **13**(4), 1362–1388.
8. M. Shahinpoor and K. J. Kim, Ionic Polymer-Metal Composites – IV. Industrial and Medical Applications, *J. Smart Mater. Struct.*, 2005, **14**(1), 197–214.
9. H.-J. Schneider, K. Kato and R. M. Strongin, Chemomechanical Polymers as Sensors and Actuators for Biological and Medicinal Applications, *Sensors*, 2007, **7**, 1578–1611.
10. H.-J. Schneider and K. Kato, Chemomechanical Polymers, in *Intelligent Materials*, Royal Society of Chemistry, Cambridge, ch. 4, 2008, pp. 100–120.
11. A. F. T. Mak and S. Sun, Intelligent Chitosan-based Hydrogels as Multifunctional Materials, in *Intelligent Materials*, Royal Society of Chemistry, Cambridge, ch. 18, 2008, pp. 447–461.
12. M. Hess, R. G. Jones, J. Kahovec, T. Kitayama, P. Kratochvil, P. Kubis, W. Mormann, R. F. T. Stepto, D. Tabak, J. Vohlidal and E. S. Wilkes, Terminology of Polymers Containing Ionizable or Ionic Groups and Polymers Containing Ions, *Pure Appl. Chem.*, 2006, **78**(11), 2067–2074.
13. K. J. Kim and M. Shahinpoor, A Novel Method of Manufacturing Three-Dimensional Ionic Polymer-Metal Composites (IPM's) Biomimetic Sensors, Actuators and Artificial Muscle, *Polymer*, 2002, **43**(3), 797–802.

14. M. Shahinpoor and K. J. Kim, Novel Ionic Polymer-Metal Composites Equipped with Physically-Loaded Particulate Electrode As Biomimetic Sensors, Actuators and Artificial Muscles, *Sens. Actuators, A*, 2002, **96**(2/3), 125–132.
15. M. Shahinpoor and K. J. Kim, A Solid-State Soft Actuator Exhibiting Large Electromechanical Effect, *Appl. Phys. Lett.*, 2002, **80**(18), 3445–3447.
16. M. Shahinpoor, Ionic Polymer-Conductor Composites As Biomimetic Sensors, Robotic Actuators and Artificial Muscles-A Review, *Electrochim. Acta*, 2003, **48**(14–16), 2343–2353.
17. S.-F. Liu and K. Schmidt-Rohr, High-Resolution Solid-State ^{13}C NMR of Fluoropolymers, *Macromolecules*, 2001, **34**, 8416–8418.
18. T. D. Gierke and W. Y. Hsu, The cluster-network model of ion clustering in perfluorosulfonated membranes, in *Perfluorinated Ionomer Membranes*, ed. A. Eisenberg and H. L. Yeager, ACS, Washington, DC, 1982, pp. 283–307.
19. T. D. Gierke, G. E. Munn and F. C. Wilson, Morphology of perfluorosulfonated membrane products—wide-angle and small-angle X-ray studies, *ACS Symp. Ser.*, 1982, **180**, 195–216.
20. M. Shahinpoor, Smart Thin Sheet Batteries Made With Ionic Polymer-Metal Composites (Ionic polymeric nanocomposites), IMECE2004-60954, *Proceedings of ASME-IMECE2004, 2004 ASME International Mechanical Engineering Congress and RD&D Exposition*, November 13–19, 2004, Anaheim, California, 2004.
21. M. Shahinpoor, A New Effect in Ionic Polymeric Gels: The Ionic "Flexoelectric Effect", *Proc. SPIE 1995 North American Conference on Smart Structures and Materials*, February 28–March 2, 1995, San Diego, California, vol. 2441, paper no. 05, pp. 42–53, 1995.
22. M. Shahinpoor, The Ionic Flexogelectric Effect, *Proc. 1996 Third International Conference on Intelligent Materials, ICIM'96, and Third European Conference on Smart Structures and Materials*, pp. 1006–1011, June 1996, Lyon, France, 1996.
23. K. Sadeghipour, R. Salomon and S. Neogi, Development of A Novel Electrochemically Active Membrane and Smart Material based Vibration Sensor/ damper, *Smart Mater. Struct.*, 1992, **1**, 172–179.
24. P. G. de Gennes, K. Okumura, M. Shahinpoor and K. J. Kim, Mechanoelectric Effects in Ionic Gels, *Europhys. Lett.*, 2003, **50**(4), 513–518.
25. K. Asaka and K. Oguro, Bending of Polyelectrolyte Membrane Platinum Composites by Electric Stimuli, Part II. Response Kinetics, *J. Electroanal. Chem.*, 2000, **480**, 186–198.
26. M. Shahinpoor and M. Mojarrad, Ionic Polymer Sensors and Actuators, *U. S. Pat. Off.*, Patent No. 6, 475,639, 2002.
27. R. Lumia and M. Shahinpoor, Microgripper design using electro-active polymers, *Proc. SPIE Smart Materials and Structures Conference*, March 1–5, 1999, New Port Beach, California, Publication No. SPIE 3669-30, pp. 322–329, 1999.
28. R. Ujwal Deole and M. S. Lumia, Design and Test of IPMC Artificial Muscle Microgripper, *Proceedings of the Third World Congress On Biomimetics, Artificial Muscle and Nano-Bio (Biomimetics and Nano-Bio 2006)*, May 25–28, 2006, Lausanne, Switzerland, 2006.
29. R. Lumia and M. Shahinpoor, IPMC Microgripper Research and Development, *Proceedings of the 4th International Congress on Biomimetics, Artificial Muscles and Nano-Bio 2007*, Cartagena, Spain, Europe, November 6–8, 2007.

18 Review of Smart Ionic Liquids

Ali Eftekhari

UK
Email: eftekhari@elchem.org

18.1 Introduction

Ionic solids such as sodium chloride (table salt) have been known for centuries. To eliminate the role of a solvent, particularly in the realm of electrochemistry, molten salts became of interest. The very first examples were the groundbreaking endeavors of Sir Humphry Davy in the synthesis of alkali metals *via* electrolysis, which he discovered (isolated to be precise) using this approach. However, a high temperature is required in this process as the ionic bonds are strong, for example, the electrolysis of sodium chloride should be conducted at a temperature higher than 801 °C. Since high temperatures are not technologically favorable, the melting point of such ionic solids can be reduced by weakening the ionic bonds in eutectic mixtures. One of the very first examples of this is in the pioneering work of Charles Martin Hall in the synthesis of aluminum, which is still the dominant approach for the exploitation of metallic aluminum. The high melting point of ionic liquids is due to the close arrangement of highly charged ions within the lattice. For instance, the sodium and chlorine are small atoms, and are closely placed in a cubic lattice structure meaning that high energy is required to separate the anions and the cations.

If the size of these ions is increased (in a sense reducing the charge density), then the melting point decreases. However, there is not

Fundamentals of Smart Materials
Edited by Mohsen Shahinpoor
© The Royal Society of Chemistry 2020
Published by the Royal Society of Chemistry, www.rsc.org

much flexibility to do so using elements, therefore, complexes are used. Although these ions can be inorganic, there are numerous organic choices. For the same reason, we have an unlimited number of organic compounds. In other words, organic–ionic compounds cannot have a high melting point because of their large and often asymmetric ions. Hence, a large number of organic–ionic compounds tend to be liquid at room temperature. Owing to both the fundamental and practical potentials of ionic compounds, which are liquid at ambient temperatures, ionic liquids (ILs) have recently attracted considerable attention. This growing attention can be seen in the number of publications in the scientific literature. Searching for the keyword "ionic liquid" in scholarly databases, in 1997, only 30 papers were published, but this number had increased to 2280 by 2007. However, it should be kept in mind that part of this massive increase is due to the terminology used. In the last ten years (2007–2017), this number was doubled to reach 5808. The emerging interest was on potential applications, and thus, the research quickly became practical rather than fundamental studies of ILs.

Despite the name, ILs are not very liquid. The definition of ILs is that they are ionic compounds with a melting point below 100 °C. This threshold surprises some people, but there was a reason for this choice. Almost all ILs form supercooled liquids below their melting points. In other words, many ILs with a melting point in the range of 25–100 °C are indeed a liquid at room temperature. However, it is evident that molten salt is a viscous liquid. Since molten salts are solvent-free media for various reactions, the first impression of ILs was that they are excellent candidates for room-temperature solvent-free media. However, the hype surrounding ILs as the solvents of the future (to replace classic solvents) did not last for long, as ILs cannot compete with classic solvents for two key reasons:

(i) Their viscosity is high due to ionic interactions and ion agglomeration, and thus, the liquidity is not at the level of molecular solvents.
(ii) ILs are expensive, not because of difficulty in synthesis, but the cost of purification. Owing to the charged nature of ILs, they are prone to contamination.

Notwithstanding, these problems did not hinder the advancement of ILs, but shifted the attention towards more practical applications. For instance, ILs exhibit attractive properties as smart materials. In such applications, the high viscosity is no longer an issue since it can be an advantage in some cases. This chapter briefly summarises the emerging applications of ILs as smart materials.

18.2 Polymerized Ionic Liquids

An interesting feature of ILs is their capability of forming a polymer chain by fixing one ion on the polymer moiety. This feature provides a rare opportunity for controlling the material properties and its ion mobility at the cost of sacrificing half the ionicity. Polymerised ILs, polymeric ILs, or poly(IL)s can all be abbreviated as PIL. The PIL conductivity is controlled by the size of the mobile ion and charge distribution. Therefore, polycations with smaller anions possess a higher conductivity compared with large asymmetrical anions.

PILs can undergo a reversible phase transition as a response to temperature. This feature is utilized in fabricating actuators. On the other hand, the thermo-responsive behavior of PILs can be used for designing membranes, the performance of which can be controlled by temperature. A detailed review of PILs from synthesis to properties can be found in the literature.[1]

18.3 Stimuli-responsive Behaviour

18.3.1 Switchable Hydrophilicity

ILs can generally be categorized as hydrophobic and hydrophilic, but this categorization is not absolute since the ionic nature of ILs is heavily controlled by temperature. This temperature control of ions means that the miscibility of a hydrophilic IL depends on the temperature. This phenomenon, which was first reported in 1998,[2] has been employed for designing thermo-responsive devices ever since. In this case, the transition between two phases and one phase occurs at the so-called critical solution temperature (CST).[3] Depending on the IL architecture, the thermo-responsive miscibility is proportional to the temperature or *vice versa*. If the solubility is increased by increasing the temperature, the upper CST (UCST) defines the lowest temperature for the formation of one phase. In the opposite case of lower CST (LCST), the miscibility is reduced by increasing the temperature. This feature is used for controlling the catalytic reactions.[4] In practice, most ILs show UCST behavior.

Owing to the flexibility of the IL architecture, their hydrophobic/hydrophilic nature can be easily controlled by various stimuli.

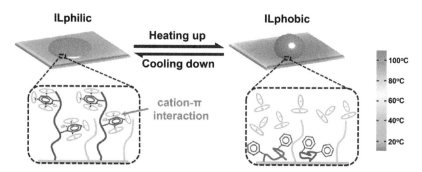

Figure 18.1 Switchable wettability of a surface designed according to the thermo-responsive properties of an IL.
Reproduced from ref. 18 with permission from the Royal Society of Chemistry.

A simple application of this phenomenon is to control the wettability, which can be beneficial in various systems. Again, the temperature can adjust the arrangement of ions, as depicted in Figure 18.1.

18.3.2 Swelling

ILs, in particular PILs, can uptake various materials within their architecture. This results in massive swelling, which can be utilized for mechanical responses in various systems from actuators to electrochemical energy storage and conversation. In the latter application, the polyelectrolytes prepared *via* this approach have a flexible structure associated with self-healing, resulting in a defect-free architecture.[5] This gel-like electrolyte is something in-between the classic liquid and solid electrolytes and gains the advantages of both types. The repulsive interaction of the inserting species and the polymer chains causes their separation, but since the polymer matrix is flexible, the polymer chains surround the uptaken species. As a result, the volume expansion is accompanied by a uniform re-arrangement of the polymer matrix without breaking the chain connections.

The swelling behavior of PILs provides a flexible medium for catalytic reactions as an analog to classic homogenous reactions.[6] Heterogeneous catalysis is usually slow because of poor dispersion of the solid catalyst within the liquid medium. However, the swelling PIL is indeed a matrix for the uniform distribution of the catalyst particles over the polymer chains, while the reactants can easily enter the medium. In other words, the swelling PIL is a smart medium, the volume of which is controlled by the reaction requirements.

18.3.3 Cross-linking Membranes

It is evident that PILs are good candidates as membranes due to the intrinsic ionic conductivity of the polymer matrix. PILs are the only members of the family of ionic polymers or polyelectrolytes that have partial ionicity. The latter have been used as commercially available ionic membranes for decades, although they are too expensive for most applications. Nanoporous PIL membranes can be easily designed *via* the arrangement of the polymer chains, particularly in a copolymer with a non-ionic polymer part. However, the interesting feature of PIL membranes is the stimuli-responsive nature of the ionic polymer chains. The simplest design is forming a copolymer in which the interaction between the polymer chains can be controlled by various stimuli thermally, optically, or chemically.[7–9] Figure 18.2 depicts how temperature and pH can subtly control the porosity and permeability of a membrane.

18.3.4 Mechanical Response

In classic actuators, the stimuli cause a mechanical response, which is normally in the form of bending. The simplest example is a bimetal thermostatic in which two metals with different thermal expansion coefficients are welded, and the strip bends towards the metal with a

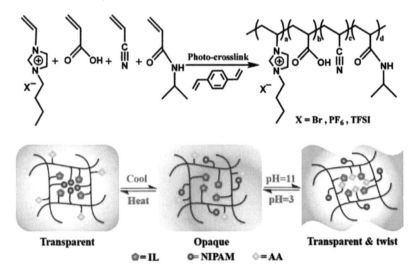

Figure 18.2 Reaction scheme for the preparation of PIL-based membranes *via in situ* photo-crosslinking and their mechanism for thermo- and pH-responsive behavior.
Reproduced from ref. 7 with permission from the Royal Society of Chemistry.

Figure 18.3 Shape transformation of the flat membrane *via* folding and recovery upon contact with water. The overall size of the flat membrane is 8.5×8.5 cm.
Reproduced from ref. 17 with permission from the American Chemical Society, Copyright 2017.

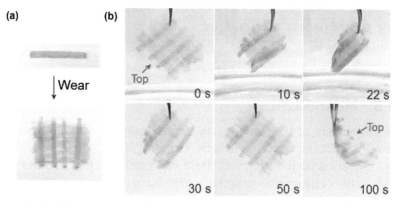

Figure 18.4 (a) A smart textile (size: 2.5×2 cm) woven from PIL/tissue paper membrane strips. (b) Actuation of the responsive textile in acetone vapor and air.
Reproduced from ref. 17 with permission from the American Chemical Society, Copyright 2017.

higher coefficient upon heating. The mechanical movement of classic actuators is simple. However, the polymer matrix of the PIL can be altered three-dimensionally. Hence, an actuator can be designed for 3D mechanical responses (Figure 18.3).

On the other hand, a PIL can be easily cast into various shapes for building complicated devices. Figure 18.4 displays simple fabrications of a series of shapes using a PIL paper with an actuating response to

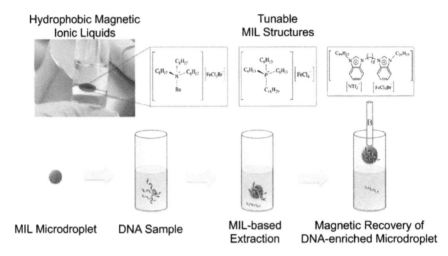

Hydrophobic Magnetic Ionic Liquids

Tunable MIL Structures

MIL Microdroplet   DNA Sample   MIL-based Extraction   Magnetic Recovery of DNA-enriched Microdroplet

Figure 18.5 Extracting DNA from a solution with the aid of a magnetic IL. Reproduced from ref. 12 with permission from the American Chemical Society Copyright 2015.

acetone vapor. All of the devices seen in Figures 18.5 and 18.6 are just simple tissue papers modified by a PIL, which reveals the simplicity of utilizing these smart materials for building actuators with complex responses.

18.3.5 Magnetic ILs

The ionic nature of ILs provides an opportunity for entanglement with magnetic particles.[10,11] This is indeed a simple design, but controlling a bunch of ions with a magnetic field can open the gates to numerous applications. Therefore, magnetic ILs have recently attracted considerable attention. For example, the ions of an IL can interact with DNA,[12] and if the IL is immobilised onto a magnetic particle, it is possible to extract DNA from a solution, as illustrated in Figure 18.5. This approach can be employed for various applications, particularly for purification purposes. As stated before, the ionic nature of ILs makes them prone to contamination. This feature, which is the prime problem in the purification of ILs, can be used for purifying other materials, such as water.[13] In this case, the IL collects the contamination, and the magnetic particle is separated by a magnetic field.

In fact, ILs are the active materials, which are carried by the core magnetic particle. Although this design can be used for any material by coating with magnetic particles, the advantage of ILs is their ionic interaction with the core magnetic particle and the environment.

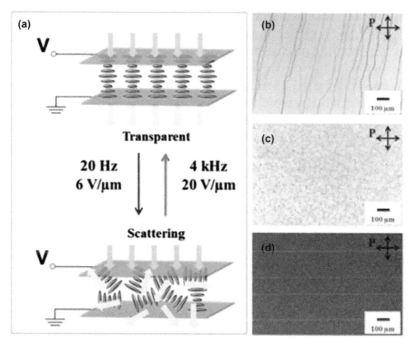

Figure 18.6 Schematic diagrams of switching between the transparent state and scattering state (green and yellow arrows represent the incident and output light, respectively). (a) POM images of a CILC cell (b) as fabricated, and as a (c) P texture driven by a high-frequency electric field of (4 kHz, 20 V μm^{-1}), and (d) FC texture driven by a low-frequency electric field of (20 Hz, 6 V μm^{-1}). The scale bar is 100 μm. The crossed arrows represent the polarizer (P) and analyzer in the POM.
Reproduced from ref. 20, https://doi.org/10.1109/JPHOT.2017. 2653862, under the terms of the CC BY 3.0 licence, https:// creativecommons.org/licenses/by/3.0/.

In other words, this is not a hard coating, and in response to stimuli, the ILs can deliver different responses, *i.e.* all of the properties described here can be used while the same material is carried by a magnetic particle.

18.3.6 Ionic Liquid Crystals

Liquid crystals (LCs) are classic examples of smart materials owing to their flexible and responsive architecture. ILs can also form LCs, which are then called ionic liquid crystals (ILCs). This class of relatively new materials has all the responsive properties of LCs plus the exceptional advantage of ionicity. This observation provides a rare opportunity when the electric signal is of importance, or ionic transport is required. For example, ILCs can provide well-guided channels for the diffusion of charge carriers in batteries.[14] By subtly designing the ILC architecture, it is possible to improve the battery performance while reducing the electrolyte viscosity, which is the main reason for the harmful leakage of batteries. The attractive point is that the viscosity, which was the principal disadvantage of ILs as electrolytes, can be an advantage for the very same application by smart design of the ILC architecture.

Smart windows for controlling the light coming into a building have an emerging market, and thus, are subject to practical research. Among various approaches, rearranging the architecture of a polymer within the LC matrix using an electrical signal is very handy. The problem is the point of stability, as usually electricity should be continuously supplied to preserve the transparent mode. ILCs provide better versatility for designing this changeable architecture.[15] Figure 18.6 illustrates how the arrangement of an ILC changes upon applying an electric field.

PILs can also be employed for designing smart windows, as the CST behavior of the IL monomer may also be retained in the resulting polymer. In this case, the polymer chain can be stimulated by several external signals. The electrical signal not only affects the charge distribution but can also chemically (to be precise electrochemically) adjust the ions to induce transparency or *vice versa*.[16]

Homework Problems

Homework Problem 18.1

Considering a series of UCST ILs: $[C_4C_1IM][BF_4]$, $[C_6C_1IM][BF_4]$, and $[C_8C_1IM][BF_4]$, estimate the order of CST values for these ILs. IM stands for imidazolium.

Homework Problem 18.2

Explain the thermoresponsive behavior illustrated in Figure 18.7. How can we build a smart device based on this system?

Homework Problem 18.3

Figure 18.8 shows a mixture of IL/water. Draw the solubility diagram for this IL.

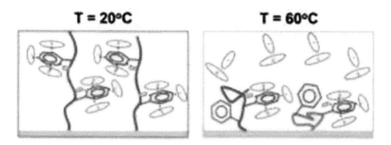

Figure 18.7 Thermoresponsive behavior.
Reproduced from ref. 18 with permission from the Royal Society of Chemistry.

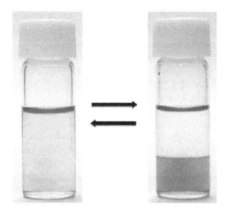

Figure 18.8 A mixture of IL/water.
Reproduced with permission from ref. 19, https://doi.org/10. 1071/CH11278, with permission from CSIRO Publishing Copyright 2011.

References

1. A. Eftekhari and T. Saito, Synthesis And Properties Of Polymerized Ionic Liquids, *Euro. Polym. J.*, 2017, **90**, 245–272.
2. J. E. L. Dullius, P. A. Z. Suarez, S. Einloft, R. F. de Souza, J. Dupont, J. Fischer and A. De Cian, Selective Catalytic Hydrodimerization Of 1,3-Butadiene By Palladium Compounds Dissolved In Ionic Liquids, *Organometallics*, 1998, **17**, 815–819.
3. D. Dupont, D. Depuydt and K. Binnemans, Overview Of The Effect Of Salts On Biphasic Ionic Liquid/Water Solvent Extraction Systems: Anion Exchange, Mutual Solubility, And Thermomorphic Properties, *J. Phys. Chem. B*, 2015, **119**, 6747–6757.
4. Y. Qiao, W. Ma, N. Theyssen, C. Chen and Z. Hou, Temperature-Responsive Ionic Liquids: Fundamental Behaviors And Catalytic Applications, *Chem. Rev.*, 2017, **117**, 6881–6928.
5. M. Yoshitake, Y. Kamiyama, K. Nishi, N. Yoshimoto, M. Morita, T. Sakai and K. Fujii, Defect-free Network Formation And Swelling Behavior In Ionic Liquid-based Electrolytes Of Tetra-arm Polymers Synthesized Using A Michael Addition Reaction, *Phys. Chem. Chem. Phys.*, 2017, **19**, 29984–29990.

6. Y. Zhang, Y. Zhang, B. Chen, L. Qin and G. Gao, Swelling Poly (Ionic Liquid)s Heterogeneous Catalysts That Are Superior To Homogeneous Catalyst For Ethylene Carbonate Transformation, *ChemistrySelect*, 2017, **2**, 9443–9449.
7. F. Chen, J. Guo, D. Xu and F. Yan, Thermo- And PH-Responsive Poly(Ionic Liquid) Membranes, *Polym. Chem.*, 2016, 7, 1330–1336.
8. X. Zhang, J. Zhou, R. Wei, W. Zhao, S. Sun and C. Zhao, Design Of Anion Species/ strength Responsive Membranes Via In-situ Cross-linked Copolymerization Of Ionic Liquids, *J. Membr. Sci.*, 2017, **535**, 158–167.
9. J. Li, J. Zhao, W. Wu, J. Liang, J. Guo, H. Zhou and L. Liang, Temperature And Anion Responsive Self-assembly Of Ionic Liquid Block Copolymers Coating Gold Nanoparticles, *Front. Mater. Sci.*, 2016, **10**, 178–186.
10. K. D. Clark, O. Nacham, J. A. Purslow, S. A. Pierson and J. L. Anderson, Magnetic Ionic Liquids In Analytical Chemistry: A Review, *Anal. Chim. Acta*, 2016, **934**, 9–21.
11. A. Joseph, G. Żyła, V. I. Thomas, P. R. Nair, A. Padmanabhan and S. Mathew, Paramagnetic Ionic Liquids For Advanced Applications: A Review, *J. Mol. Liq.*, 2016, **218**, 319–331.
12. K. D. Clark, O. Nacham, H. Yu, T. Li, M. M. Yamsek, D. R. Ronning and J. L. Anderson, Extraction Of DNA By Magnetic Ionic Liquids: Tunable Solvents For Rapid And Selective DNA Analysis, *Anal. Chem.*, 2015, **87**, 1552–1559.
13. H. Yang, H. Zhang, J. Peng, Y. Zhang, G. Du and Y. Fang, Smart Magnetic Ionic Liquid-based Pickering Emulsions Stabilized By Amphiphilic Fe3O4 Nano-particles: Highly Efficient Extraction Systems For Water Purification, *J. Colloid Interface Sci.*, 2017, **485**, 213–222.
14. A. Eftekhari, Y. Liu and P. Chen, Different Roles Of Ionic Liquids In Lithium Batteries, *J. Power Sources*, 2016, **334**, 221–239.
15. H. Bai, B. Hao, F. Walsh, W. Flynn, B. Gludovatz, G. Bernd, B. Delattre, D. Benjamin, C. Huang, H. Caili, Y. Chen, C. Yuan, A. P. Tomsia and R. O. Ritchie, Bioinspired Hydroxyapatite/Poly(methyl Methacrylate) Composite With A Nacre-Mimetic Architecture By A Bidirectional Freezing Method, *Adv. Mater.*, 2016, **28**, 50–56.
16. K. Zhang, M. Zhang, X. Feng, M. A. Hempenius and G. J. Vancso, Switching Light Transmittance By Responsive Organometallic Poly(Ionic Liquid)s: Control By Cross Talk Of Thermal And Redox Stimuli, *Adv. Funct. Mater.*, 2017, **27**, 1702784.
17. H. Lin, J. Gong, H. Miao, R. Guterman, H. Song, Q. Zhao, J. W. C. Dunlop and J. Yuan, Flexible And Actuating Nanoporous Poly(Ionic Liquid)–Paper-Based Hybrid Membranes, *ACS Appl. Mater. Interfaces*, 2017, **9**, 15148–15155.
18. L. Chang, H. Liu, Y. Ding, J. Zhang, L. Li, X. Zhang, M. Liu, L. Jiang and A. Smart, Surface With Switchable Wettability By An Ionic Liquid, *Nanoscale*, 2017, **9**, 5822–5827.
19. Y. Kohno, H. Arai, S. Saita and H. Ohno, Material Design Of Ionic Liquids To Show Temperature-sensitive LCST-type Phase Transition After Mixing With Water, *Aust. J. Chem.*, 2011, **64**, 1560–1567.
20. Z. Lan, Y. Li, H. Dai and D. Luo, *IEEE Photonics*, 2019, **1027**(9), 2200307.

19 Review of Conductive Polymers as Smart Materials

Mohsen Shahinpoor

Mechanical Engineering Dept., University of Maine, USA
Email: shah@maine.edu

19.1 Introduction

There are currently a fairly large number of conducting polymers or synthetic conductors that are being used industrially or medically. Some of the basic conducting polymers are polypyrrole, polyaniline, polythiophene, polyphenyl vinylene, polyacetylene, *etc.*, which can be manufactured *via* chemical or electrochemical oxidation and reduction (REDOX) procedures. Conductive polymers with the ability to conduct electrical charges in addition to being flexible, optically active and not difficult to synthesize present a tremendous opportunity for the industrial and medical applications of conductive polymers.

Pioneering work[1–4] on conductive polymers reported the observation that the conductivity of polyacetylene increases by millions of times when it is oxidized by "doping" with iodine vapor. Only through this "doping" does the conductivity of the polymer increased. Conductive polymers can conduct electrical charge due to charges being able to jump between the molecular chains of their molecular network.[1–4] Conductive polymer molecular structures possess both single and double chemical bonds, which enhance charge transfer.[1–4]

The discovery of polyacetylene and its high conductivity led to the field of organic conductive polymers or synthetic metals and

Fundamentals of Smart Materials
Edited by Mohsen Shahinpoor
© The Royal Society of Chemistry 2020
Published by the Royal Society of Chemistry, www.rsc.org

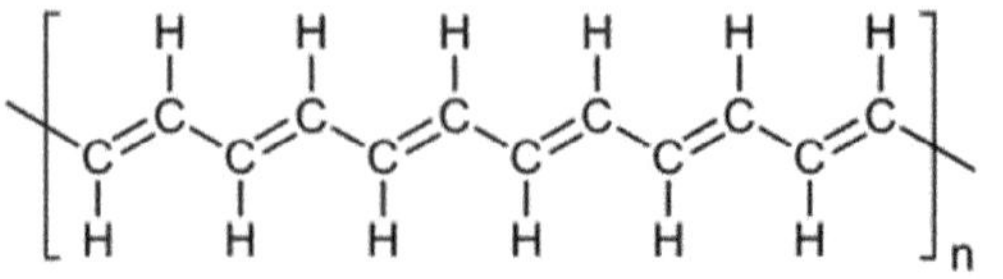

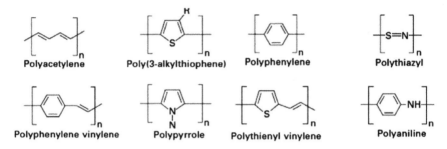

Figure 19.1   The chemical formula of polyacetylene.

Polyacetylene   Poly(3-alkylthiophene)   Polyphenylene   Polythiazyl

Polyphenylene vinylene   Polypyrrole   Polythienyl vinylene   Polyaniline

Figure 19.2   Chemical formulas of leading conductive polymers.

eventually this discovery was recognized by the Nobel Prize in Chemistry in 2000, which was awarded to Hideki Shirakawa, Alan Heeger, and Alan MacDiarmid.[1–4] Figure 19.1 depicts the chemical formula of polyacetylene.

Polyacetylene has a high conductivity of more than $100\,000\,\mathrm{S\,cm^{-1}}$ compared to those of polypyrrole at $600\,\mathrm{S\,cm^{-1}}$, polythiophene at $200\,\mathrm{S\,cm^{-1}}$, polyphenylene at $500\,\mathrm{S\,cm^{-1}}$, and polyaniline at $10\,\mathrm{S\,cm^{-1}}$.

Figure 19.2 depicts the chemical formulas of some other important conductive polymers.

These discoveries led to increased interest in the use of soft organic compounds in microelectronics. Generally, the prevailing anions within a conductive polymer interchange during the redox operation resulting in an increase in volume during oxidation and a decreasing in volume during reduction. Electrochemical processes involving conducting polymers include the flow of hydrated ions and loose water molecules across the molecular network and re-arrangements of double bonds along the chains. The changes in volume caused by internal ionic migration can induce large bending and twisting deformation in conductive polymers. Conductive polymers are also capable of sensing, thus mimicking biological muscles. For further information, see ref. 5–12.

According to Balint, Cassidy and Cartmell,[13] conductive polymers will revolutionize the world of tissue engineering. In their review article, thorough information is provided on the most commonly used conductive polymers, their conductivity, biocompatibility, biomolecule doping, synthesis, and drug delivery applications. According to ref. 13, there are over 25 conductive polymers, the abbreviations of which are listed in Box 19.1.

Box 19.1 List of typical conductive polymers and their abbreviations

Polypyrrole (PPy), Polyaniline (PANI), Poly(3,4-ethylene dioxythiophene) (PEDT, PEDOT), Polythiophene (PTh), Polythiophene-vinylene (PTh-V), Poly(2,5-thienylenevinylene) (PTV), Poly(3-alkyl thiophene) (PAT), Poly-(*p*-phenylene) (PPP), Poly-*p*-phenylene–sulfide (PPS), Poly(*p*-phenylene vinylene) (PPV), Poly(*p*-phenylene–terephthalamide) (PPTA), Polyacetylene (PAc), Poly(isothianaphthene) (PITN), Poly(α-naphthylamine) (PNA), Poly-azulene (PAZ), polyfuran (PFu), Polyisoprene (PIP), Polybutadiene (PBD), Poly(3-octylthiophnene-3-methylthiophene) (POTMT), Poly(*p*-phenylene–terephthalamide) (PPTA)

For a detailed description of these conductive polymers and their applications, see ref. 13.

19.2 Conductivity of Conductive Polymers

As described in the introduction, conductive polymers will revolutionize the world of tissue engineering.[13] However, there are already many industrial and medical applications that are related to their conductivity, flexibility, biocompatibility, biomolecule doping, sensing, synthesis, and drug delivery properties. As far as the conductivity of these polymers are concerned, note that as previously described, polyacetylene has the highest conductivity value, comparable to that of metals, of around $100\,000$ S cm^{-1} for maximum values. Compared to polyacetylene, polypyrrole has a conductivity of about 7500 S cm^{-1}, polyaniline has a maximum conductivity of 200 S cm^{-1}, polythiophene has a maximum conductivity of 1000 S cm^{-1}, poly(*p*-phenylene) has a maximum conductivity of 1000 S cm^{-1}, and poly(*p*-phenylene vinylene) has a maximum conductivity of 5000 S cm^{-1}.[13,14] These conductivities depend on the dopants used and may vary.

19.3 Electro–Chemo–Mechanical Properties

The electro–chemo–mechanical characteristics of conductive polymers are governed by the entrance and expulsion of counterions within the polymeric network, in an electrolyte. Water driven by the electrochemical reaction promotes reversible changes in the material volume. The electrochemical reactions within the network control these changes and can induce deformation and shape change if placed in an electrolyte with the insertion of appropriate charges by an imposed electric field, as described by Otero.[14]

19.4 Experimental Observations on Conductive Polymers (CPs)

Results on reversible volume change associated with the electro-chemical reactions of polypyrrole in various electrolytes have been reported by Otero *et al.*[10] Figure 19.3 depicts the reversible volume change associated with the electrochemical reactions of polypyrrole in electrolyte. Here, prevailing anions tend to interchange during the redox processes, increasing the volume during oxidation and decreasing the volume during reduction, in electrolyte.

On the other hand, when cations prevail, the conductive polymer behavior is reversed. Figure 19.4 depicts the reversible volume change associated with the electrochemical reactions of polypyrrole in electrolyte.

Here, prevailing cations tend to interchange during the redox processes, decreasing the volume (shrinking) during oxidation and increasing the volume during reduction, in electrolyte. Here, a metal/polymeric blend/solution system is formed and the polymeric blend is made up of entangled structures of a conductive polymer such as polypyrrole and a polyelectrolyte such as polyvinyl sulfonate. These are obtained by electrogeneration from a solution containing the monomer and the solvated polyelectrolyte. During oxidation or reduction, cations are expulsed or inserted from the solution and the polymer shrinks or swells, respectively. For a detailed explanation of these ionic dynamics, refer to the study by Otero.[14]

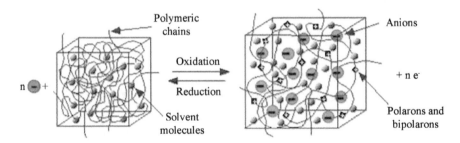

Figure 19.3   Schematic representation of the reversible volume change associated with the electrochemical reactions of polypyrrole in electrolyte.
Reproduced from ref. 12 with permission from Springer Nature, Copyright 2000.

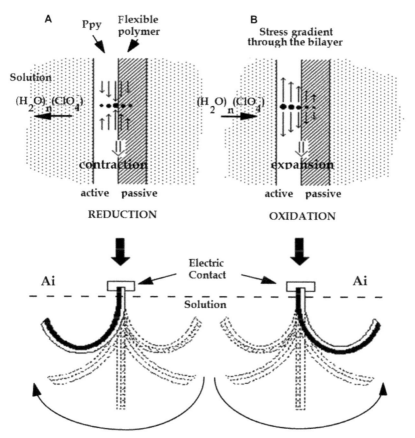

Figure 19.4 Schematic of a CP bilayer (active) adherent, flexible, and non-conducting polymer tape (passive). Schematics of the ionic interchanges between the CP and the solution during the electrochemical reactions produced by the current (A) and of the stress gradients between both layers (B) induced by the ionic interchanges and concomitant volume changes in the active layer of the CP.
Reproduced from ref. 14 with permission from Springer Nature, Copyright 1993.

19.5 Bending Structures

According to Otero,[14] a membrane of a CP with prevalent anion interchange can be electrogenerated by metal electrodes. The membrane used as an anode in an electrolyte undergoes an asymmetric change in volume; swelling the CP where anion interchange with the solution prevails and shrinking the CP where the interchange with cations prevails.

Metallic counter electrodes are then required to allow the current flow. The membranes also produce a bending movement, provided an electrolyte and a counter electrode are present to allow the flow of

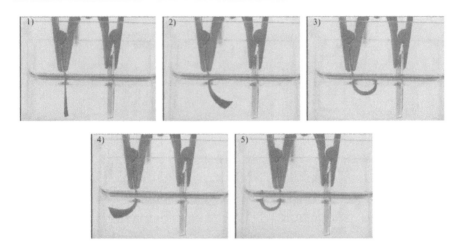

Figure 19.5   Macroscopic bending in polyelectrolyte solution during anodic
(1–3) or cathodic (4, 5) oxidation or reduction, respectively, by a
current flow through a CP/tape device. Different positions of a
bilayer muscle (1–5) showing the metallic counter electrodes
and the solution, where 1–3 shows the flow of an anodic current
of 15 mA, and 4–5 the flow of a cathodic current of −15 mA.
Reproduced from ref. 14 with permission from Springer Nature,
Copyright 1993.

the electrical current and to produce the bending movement.
When the conductive polymer film has a prevailing interchange of
anions, the oxidation causes the expansion or swelling of the CP
(Figures 19.4 and 19.5).

According to Otero,[14] one can create a triple layer conductive polymer
actuator using a polymeric tape and bonding membranes of CP to each
side. One of the CP films is connected to the working electrode output
from a potentiostat. The second CP film is connected to the counter
electrode (CE) output and short-circuited with the reference electrode
(RE). By placing the triple layer in an electrolyte solution, the structure of
the triple-layer electrochemical cell allows the current to flow through it,
as depicted in Figure 19.6. Note that now the same current flows through
the two polymeric films such that the anode swells and the cathode
shrinks, resulting in the bending of the material.

Figure 19.7 depicts the structure of the triple layer conductive
polymer actuator, as suggested by Otero.[14,15] Here, the conductive
polymer actuator has a three layer polypyrrole/nonconducting
film/polypyrrole structure. Thus, the consumed charge has two
functions in this device: when polypyrrole I is oxidized (anodic
process) the free end of the layer is deformed while polypyrrole II is
reduced (cathodic process).

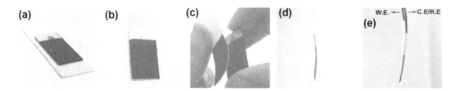

Figure 19.6 Consecutive steps (a–e) for the construction of a triple-layer muscle (CP/tape/CP) starting from electrogenerated films of a CP on metallic electrodes.
Reproduced from ref. 8 with permission from Springer Nature, Copyright 2002 Kluwer Academic Publishers.

Exercise Number 19.1: What are the possible conductive polymer actuator arrangements with electrodes?

Answer to Exercise No. 19.1: A membrane of a conductive polymer (CP) with prevalent anion interchange can be electrogenerated by metal electrodes. The membrane used as an anode in an electrolyte undergoes an asymmetric change of volume: swelling the CP where the anions interchange with the solution prevail and shrinking the CP where the interchange of cations prevails.

Metallic counter electrodes are then required to allow the current flow. The membranes also produce a bending movement provided an electrolyte, and a counter electrode is present to allow the flow of the electrical current and to produce the bending movement. When the conductive polymer film has a prevailing interchange of anions, the oxidation causes the expansion or swelling of the CP.

19.6   Fabrication and Manufacturing

According to ref. 8, 13, there are two main methods for synthesizing conductive polymers; one is "chemical" and the second is "electrochemical." In chemical synthesis, the monomer is oxidized in a polyelectrolyte solution with an oxidizing agent to create a thick film of the polymer that can be mass produced, which is appropriate for commercial applications.

According to ref. 13, most conductive polymers are prepared *via* the oxidative coupling of monocyclic precursors. Such chemical reactions involve the removal of hydrogen from an organic molecular network (dehydrogenation):

$$n\ \text{H}-[\text{X}]-\text{H} \rightarrow \text{H}-[\text{X}]_n-\text{H} + 2(n-1)\ \text{H}^+ + 2(n-1)\ \text{e}^- \qquad (19.1)$$

According to ref. 13, the dehydrogenation converts alkanes to olefins, which are active multi-functional materials (polyolefins as shape memory polymers). The chemical method of manufacturing

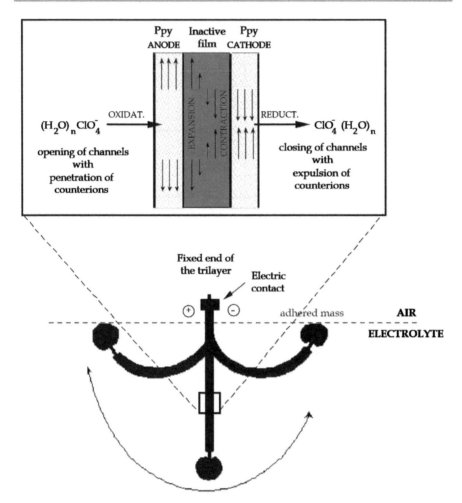

Figure 19.7 Schematic diagram of ionic interchanges, induced stress gradients and generated angular movements during current flow. Reproduced from ref. 8 with permission from Springer Nature, Copyright 2002 Kluwer Academic Publishers.

conductive polymers leads to lower conductivities, which are further highly sensitive to several factors such as the kind and purity of the solvent, the oxidant, the relative concentration of the reagents, reaction time, temperature and stirring rate. In electrochemical polymerization, a solution containing the monomer of the conductive polymer is equipped with two electrodes through which an electrical current is applied to the solution with the doping agents. The electrical current forces the monomer to oxidize, which is consequently attracted to and deposits at the anode. Electrochemical polymerization is used to synthesize the CP, through oxidation in the presence of an electrical potential. Electrochemical production of CPs is very common, especially for the main conductive polymers currently in use. It should be mentioned that electrochemical synthesis involves the use of three potent electrical techniques to produce CPs, namely galvanostatic, potentiostatic, and potentiodynamic methods. In potentiostatic polymerization, the potential is controlled, which is ideal for the manufacture of CP biosensors. On the other hand, in galvanostatic polymerization, the electrical current is controlled instead of the potential, which results in the steady-state deposition of the oxidized polymer. During potentiodynamic deposition, the polymerizing potential is cyclically varied between a low and high limit, which causes the polymer to be deposited in electrically active layers.[13]

19.7 Conclusions

This chapter presented a very brief review of CPs as smart multifunctional materials. Typical materials with their synonyms are polypyrrole (PPy), polyaniline (PANI), poly(3,4-ethylene dioxythiophene) (PEDT, PEDOT), polythiophene (PTh), polythiophene-vinylene (PTh-V), poly(2,5-thienylenevinylene) (PTV), poly(3-alkyl thiophene) (PAT), poly(*p*-phenylene) (PPP), poly-*p*-phenylene-sulfide (PPS), poly(*p*-phenylene vinylene) (PPV), poly(*p*-phenylene-terephthalamide) (PPTA), polyacetylene (PAc), poly(isothianaphthene) (PITN), poly(α-naphthylamine) (PNA), polyazulene (PAZ), polyfuran (PFu), polyisoprene (PIP), polybutadiene (PBD), poly(3-octylthiophnene-3-methylthiophene) (POTMT), poly(*p*-phenylene terephthalamide) (PPTA). It has been reported that there are already many industrial and medical applications that use CPs, as a result of their conductivity, flexibility, biocompatibility, biomolecule doping, sensing, synthesis, and drug delivery applications. New materials and greater control of their syntheses will be required to produce tailored materials for use in serious engineering applications.

Homework Problems

Homework Problem 19.1

Which conductive polymer has the highest conductivity and why?

Homework Problem 19.2

What is the most exciting biomedical application of conductive polymers?

References

1. A. Heeger, Nobel Lecture: Semiconducting and metallic polymers: The fourth generation of polymeric materials, *Rev. Mod. Phys.*, 2001, **73**(3), 681.
2. Heeger, MaDiarmid and Shirakawa, *The Nobel Prize in Chemistry 2000*, 2000, https://www.nobelprize.org/prizes/chemistry/2000/summary/.
3. H. Shirakawa, Synthesis and characterization of highly conducting polyacetylene, *Synth. Met.*, 1995, **69**, 3.
4. H. Shirakawa, E. J. Louis, A. G. MacDiarmid, C. K. Chiang and A. J. Heeger, Synthesis of Electrically Conducting Organic Polymers: Halogen Derivatives of Polyacetylene, $(CH)_x$, *J. Chem. Soc., Chem. Commun.*, 1977, **16**, 578–580.
5. *Handbook of Conducting Polymers*, ed. T. Stotheim, R. Eisenhauer and J. Reynolds. Marcel Dekker, Inc., New York, 1998.
6. *Handbook of Organic Conductive Molecules and Polymers*, ed. H. S. Nalwa, John Wiley & Sons, 1997.
7. P. Chandrasekhar, *Conducting Polymers, Fundamental, and Applications*, Kluwer Academic Publishers, Boston, 1999.
8. T. F. Otero, Conducting polymers, electrochemistry, and biomimicking processes, in *Modern Aspects of Electrochemistry*, ed. J. O'M. Bockris, R. E. White and B. E. Conway, Plenum, New York, vol. 33, 1999, pp. 307–434.
9. G. G. Wallace, G. M. Spinks, L. A. P. Kane-Maguire and P. R. Teasdale, *Conductive Electroactive Polymers. Intelligent Materials*, CRC Press, Boca Ratón, 2003.
10. *Polymer Sensors and Actuators*, ed. T. O. Otero, Y. Osada and D. E. De Rossi, Springer-Verlag, Berlin, 2000, p. 304.
11. R. H. Baughman and L. W. Shacklette, *Science and Application of Conducting Polymers*, ed. W. R. Salaneck, D. T. Clark and E. J. Samuelson, Adam Hilger, Bristol, 1991, p. 47.
12. T. F. Otero, Electrochemomechanical devices based on conducting polymers, in *Polymer Sensors and Actuators*, ed. D. de Rossi and Y. Osada, Springer-Verlag, Oxford, 2000, pp. 295–323.
13. R. Balint, N. J. Cassidy and S. H. Cartmell, Conductive polymers: Towards a smart biomaterial for tissue engineering, A Review, *Acta Biomater.*, 2014, **10**(6), 2341–2353.
14. T. F. Otero, Artifical Muscles, Sensing and Multifunctionality, in *Intelligent Materials*, ed. M. Shahinpoor and H. J. Schneider, Royal Society of Chemistry, Cambridge, ch. 6, 2008.
15. T. F. Otero and J. Rodriguez, Electrochemomechanical and Electrochemopositioning Devices: Artificial Muscles, in *Intrinsically Conducting Polymers: An Emerging Technology*, ed. M. Aldissi, NATO ASI Series (Series E: Applied Sciences), Springer, Dordrecht, vol. 246, 1993.

20 Review of Liquid Crystal Elastomers

Mohsen Shahinpoor

Mechanical Engineering Dept., University of Maine, USA
Email: shah@maine.edu

20.1 Introduction

Liquid crystal elastomers (LCEs) were first discussed by Finkelmann *et al.*[1–16] These materials can be used as robotic actuators upon undergoing a nematic–isotropic phase transition upon a temperature increase, which causes them to shrink, as described in a thorough review on these intelligent multi-functional materials by Brand and Finkelmann.[2] LCEs have been made electroactive by creating a composite material that consists of nematic LCEs and a conductive phase such as graphite or synthetic metals (conductive polymers) that are distributed within their molecular network structures.[3,4] Liquid crystal elastomers perform actuation based on certain isotropic–anisotropic phase transitions.

20.2 Brief Background on Liquid Crystals

Liquids are generally *isotropic,* meaning that their properties are uniform in all directions. However, a liquid crystalline phase is anisotropic, and thus has different properties in different directions, and yet can flow like a liquid. Liquid crystal phases possess sufficiently ordered molecular units to become highly anisotropic, and

Fundamentals of Smart Materials
Edited by Mohsen Shahinpoor
© The Royal Society of Chemistry 2020
Published by the Royal Society of Chemistry, www.rsc.org

this anisotropy is very much evident in their optical properties. It was in the mid-1960s when the Nobel laureate French physicist Pierre-Gilles de Gennes (1932–2007) developed a thorough theoretical model to describe the properties of liquid crystalline phases, in particular their ability to scatter light.

20.3 Nematic, Cholesteric and Smectic Phases of Liquid Crystals

There are three types of main liquid crystalline phases, namely nematic, smectic, and cholesteric, as shown in Figures 20.1 and 20.2.

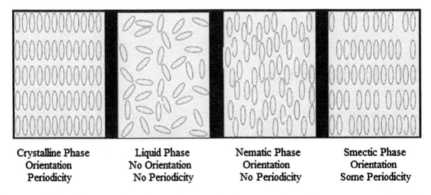

Crystalline Phase	Liquid Phase	Nematic Phase	Smectic Phase
Orientation	No Orientation	Orientation	Orientation
Periodicity	No Periodicity	No Periodicity	Some Periodicity

Figure 20.1 Conventional condensed phases of matter compared with a crystalline phase, where non-crystalline, oriented and periodic liquid crystal structures are shown.

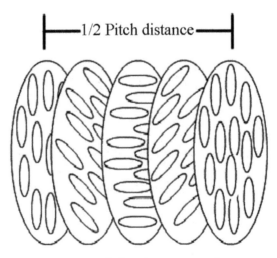

Figure 20.2 General chirality of the directors and structure of cholesteric liquid crystals.

In a *nematic phase* ("thread-like") the molecules are aligned in the same direction but are free to drift around randomly, very much like as in an ordinary liquid, where the alignment of the charged polar rod-like molecules can be changed by applying an electric field.

In a *smectic* ("soap-like") liquid crystalline phase, the molecules stack up in layers that slide upon each other, with the long molecular axes approximately perpendicular to the laminar planes of sliding. If these layers are helically arranged along the general axis, the crystalline phase is called cholesteric. If the molecules are polar, this helical twisting can be reversed by applying an external electric field at either end of the long axis. Figure 20.2 depicts the general spiral structure of a cholesteric liquid crystal.

Figure 20.3 shows a chiral molecule, where the chiral pendant structure of the molecule is indicated using a * symbol.

It was Nobel laureate Pierre de Gennes and co-workers Herbert and Kant[5,17,18] who proposed the use of liquid crystals as artificial muscles and pointed out that they could be used as multi-functional materials, and in particular, biomimetic artificial muscles. Later, Finkelmann, Brand, *et al.*[1-4] made significant contributions towards the multi-functionality of liquid crystal elastomers. It was shown how blends of liquid crystal elastomers and a conductive phase such as graphite could lead to smart materials such as electrically controllable biomimetic artificial muscles.

These materials can be synthesized *via* the hydrosilylation reaction of the monofunctional side chain mesogen (Figure 20.4) and the bifunctional liquid crystalline main chain polyether with poly(methylhydrosiloxane). Figure 20.4 depicts how the nematic co-elastomer is mechanically loaded in a liquid single-crystalline co-elastomer, a weakly cross-linked network is mechanically loaded. Under load, the hydrosilylation reaction is completed using PHMS (Polymethylhydrosiloxane) at 60 °C in the presence of platinum as a catalyst. The liquid single-crystal elastomer (LSCE) undergoes a nematic–isotropic phase transition, in a thermal bath, which is accompanied by contraction, as shown in Figure 20.5.

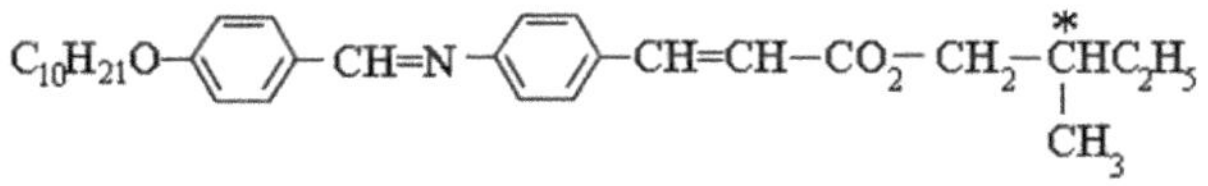

$$C_{10}H_{21}O-\langle\bigcirc\rangle-CH=N-\langle\bigcirc\rangle-CH=CH-CO_2^- CH_2-\overset{*}{C}HC_2H_5$$
$$| \atop CH_3$$

Figure 20.3 Typical chiral molecule that gives rise to helically twisted cholesteric structures.

side chain mesogen

main chain mesogen/crosslinker

PMHS

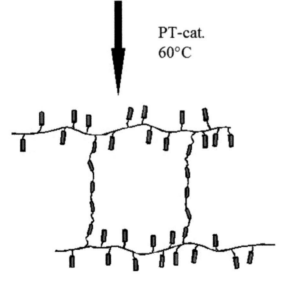

PT-cat.
60°C

Figure 20.4   Synthesis of a nematic co-elastomer using PHMS.
Reproduced from ref. 20 with permission from the Royal Society of Chemistry.

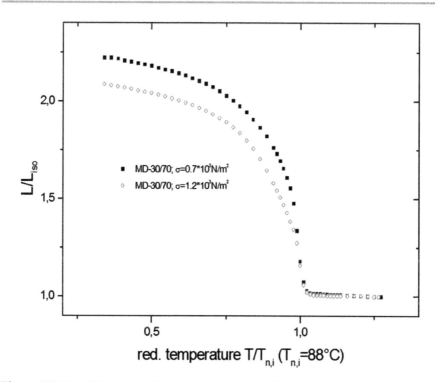

Figure 20.5 LCE contraction in a temperature bath.
Reproduced from ref. 20 with permission from The Royal Society of Chemistry.

LSCEs are liquid crystalline phase structures with preferred macroscopic molecular orientation in their molecular network. When the temperature of the LCE phase is at the phase transition temperature, the liquid crystalline phase becomes isotropic (rubber-like) and the liquid crystalline network deforms. Here, the liquid crystalline network shortens in the optical axis or molecular polar direction due to the imposed anisotropy. De Gennes and co-workers[17,18] proposed that nematic liquid crystalline networks are capable of producing artificial muscle-type dynamic behavior. It was observed[1–5,12,14,17,18] that some nematic LSCEs synthesized from nematic LC side chain polymers, are capable of producing more than 100% linear strain. The speed of this deformation (length change) is determined by the thermal conductivity of the network and not by material transport processes. At the transformation temperature at which the liquid crystalline phase becomes isotropic, like a conventional rubber, the dimensions of the network change.

As was mentioned before, electrically controllable LSCEs can be fabricated by making composites with electrically conductive media

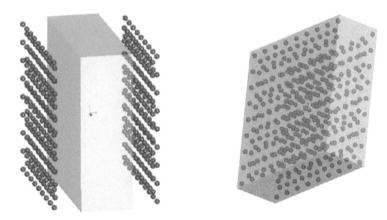

Figure 20.6 Loading of graphite powder into liquid crystal elastomer films. Reproduced from ref. 20 with permission from the Royal Society of Chemistry.

such as graphite, metal powders, or *via* the physical loading of an LCE with a conductive phase, as shown in Figure 20.6. Here, the idea was to repeat the same type of experiment as depicted above in Figure 20.5, in which the elastomer shrank upon heating in a liquid thermal bath, *via* Joule heating of an LSCE with embedded graphite fibers inside the elastomer.

In summary, co-elastomers containing network strands comprising LC side and main chain polymers exhibit exceptional thermoelastic behavior. Thus, they may be appropriate for soft artificial muscle applications.

Exercise No. 20.1 What specific chemicals and conditions are needed for the synthesis of nematic liquid crystals?

Answer to Exercise No. 20.1 The specific chemical needed for the synthesis of a nematic co-elastomer is PMHS (polymethylhydrosiloxane) in the presence of platinum as a catalyst and a temperature of 60 °C temperature, as shown in Figure 20.4.

The hydrosilylation reaction is completed under loading conditions. Macroscopically disordered LCEs result when there is no mechanical loading.

20.4 Electrically-controllable Liquid Crystal Elastomer–Graphite Composites (LCE–G)

Here, we expand upon this notion based on work by Finkelmann and Shahinpoor.[19] An LCE membrane (12 mm×24 mm×0.3 mm) is blended with layers of fine 2 micron diameter graphite powder and subjected to 100 kPa of pressure to blend the powder into a liquid crystal membrane surface. The blended LCE–G composite will become electrically conductive if the volume fraction of blended conductive graphite particles is above the percolation threshold of 35%.[19] The LCE–G composite will have an effective resistivity of about 2 ohms cm^{-1} on its surface and about 1 ohm cm^{-1} across its thickness.

Upon applying a low DC voltage (1–5 volts) to the samples for a few seconds and under a stress loading of around 10 kPa, the sample will contract quickly in around a second to about 18 mm lengthwise, with an average linear strain of about 25%. The samples generally have negligible contractions in the thickness and transverse directions.

Some more experimental methods are described in the next section.

20.5 Experimental Procedure

Since LCEs undergo phase transitions due to temperature variations, it is convenient to make an LCE electrically conductive, as discussed in the previous section. Blending or mixing LCEs with conductive fine granular materials or powders such as carbon, graphite, graphene, carbon nanotubes and possibly particles of conducting polymers such as polypyrrole or polyaniline (synthetic metals) will make then electrically conductive and enable the electric Joule heating of the LCE to induce phase transformation.

Here, the goal of these efforts is to press conducting particles inside the elastomer to enhance its conductivity and make it easier to Joule heat it. Here, we report on this notion using novel liquid crystalline co-elastomer materials.

To make an electrically nonconductive phase conductive, the uniform distribution of the conductive phase is necessary. The volume of graphite powder to the LCE film in typical experiments[19,20] was more than 53%, well over the percolation threshold (33%) needed to make the elastomer conductive. Figure 20.7 depicts such a nematic–isotropic phase transition, based on the experimental results of Finkelmann *et al.*[16]

20.6 Modeling and Constitutive Equations

The reader is referred to work by Shahinpoor *et al.*[15,19–26] for modeling and constitutive equations on LCEs. Essentially, the modeling of

Elastomer	MU_{SK}/MU_{HK}	crosslinker %	$\overline{DP}$	T_g [°C]	$T_{n,i}$ [°C]	swelling degree	swelling anisotropy
MD-20-13	80/20	1.9	13	7	89	6.9	1.8
MD-28-13	**72/28**	**2.9**	**13**	**8**	**86**	**6.3**	**2.9**
MD-46-20	54/46	4.0	20	13	95	6.5	4.1
MD-57-26	43/57	4.6	26	19	96	6.8	4.5

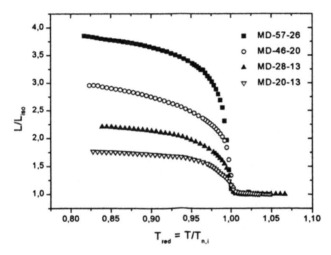

Figure 20.7   Contraction of LCEs with temperature.[15,19,20]

LCEs is based on the interaction of the internal directors and their initial configurations. Based on these directors, both molecular and continuum modeling of the constitutive equations have been carried out and the results are available in the literature.[15,19–26] Figures 20.7 and 20.8 depict the Joule heating of two different LCEs subjected to different loads.

20.7   Conclusions

Liquid crystals and LCEs were introduced in this chapter as multi-functional smart materials. LCEs comprising polar molecular network strands of liquid crystal side and main chain polymers have been reported to exhibit exceptional thermoelastic behavior. Thus, LCEs may be suitable to use as soft biomimetic artificial muscles for actuation purposes. Electrically conducting composites of LCEs can be manufactured using a dispersed uniform phase of an

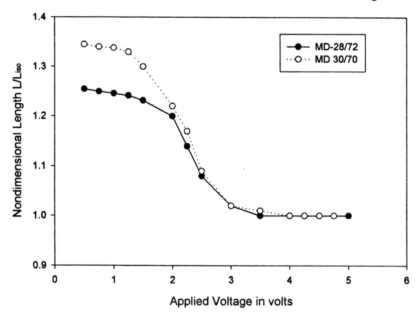

Figure 20.8 Contraction of an LCE as a function of voltage under a tensile load of 10 kPA.[15,19,20]

electrically conducting granular (powder-like) material such as carbon, graphite, graphene or carbon nanotubes. Fabrication methods and experimental results on a new class of LCE–graphite (LCE–G) composites as electrically-controllable biomimetic artificial muscles were described.

Homework Problems

Homework Problem 20.1

Describe the structural differences between nematic, cholesteric, and smectic liquid crystals in terms of orientations of internal directors.

Homework Problem 20.2

Describe a possible manufacturing paradigm of making an electrically conductive liquid crystal elastomer as a composite with a conductive phase.

References

1. H. R. Brand and H. Finkelmann, An overview of research in liquid crystalline elastomers with a complete set of references up to 1997, Physical Properties of Liquid Crystalline Elastomers, in *Handbook of Liquid Crystals Vol. 3: High Molecular Weight Liquid Crystals*, ed. D. Demus, J. Goodby, G. W. Gray, H.-W. Spiess and V. Vill, Wiley-VCH, Weinheim, 1998, p. 277.
2. H. Finkelmann, H. J. Kock and G. Rehage, Investigations on Liquid Crystalline Siloxanes: 3. Liquid crystalline elastomers – a new type of liquid crystalline material, *Makromol. Chem., Rapid Commun.*, 1981, **2**, 317.
3. J. Küpfer and H. Finkelmann, Nematic Liquid Crystal Elastomer, *Macromol. Chem., Rapid Commun.*, 1991, **12**, 717–726.
4. H. Finkelmann and H. R. Brand, Liquid Crystalline Elastomers - A Class of Materials with Novel Properties, *Trends Polym. Sci.*, 1994, **2**, 222.
5. J. Küpfer and H. Finkelmann, Nematic Single Liquid Crystal Elastomers, *Makromol. Chem.*, 1994, **195**, 1353.
6. H. R. Brandt and K. Kawasaki, On the macroscopic consequences of frozen order in liquid single crystal elastomers, *Macromol. Rapid Commun.*, 1994, **15**, 251.
7. H. R. Brandt and H. Pleiner, Electrohydrodynamics of nematic liquid crystalline elastomers, *Physica*, 1994, **A208**, 359.
8. W. Kaufhold, W. Finkelmann and H. R. Brand, Nematic Elastomers: 1. Effect of the spacer length on the mechanical coupling between network anisotropy and nematic order, *Makromol. Chem.*, 1991, **192**, 2555.
9. S. Disch, S. Finkelmann, H. Ringsdorf and P. Schumacher, Macroscopically ordered discotic networks, *Macromolecular*, 1995, **28**, 2424.
10. J. Küpfer, E. Nishikawa and H. Finkelmann, Densely cross-linked liquid single crystal elastomers, *Polym. Adv. Technol.*, 1994, **5**, 110.
11. E. Nishikawa, Smektische Einkristall-Elastomere, PhD Thesis, Albert-Ludwigs-Universität Freiburg im Breisgau, 1997.
12. G. H. F. Bergmann, H. Finkelmann, V. Percec and M. Zhao, *Macromol. Rapid Commun.*, 1997, **18**, 353.
13. G. H. F. Bergmann, Flüssigkristalline Hauptketten elastomere: Synthese, Characterisierung und Untersuchungen zu mechanishen, thermischen und Orientierungseigenschaften, PhD thesis, Albert-Ludwigs-Universität Freiburg im Breisgau, 1998.
14. G. Bergmann and H. Finkelmann, *Macromol. Chem., Rapid Commun.*, 1997, **18**, 353–359.
15. M. Shahinpoor, Electrically-Activated Artificial Muscles Made With Liquid Crystal Elastomers, *Proceedings of SPIE 7th International Symposium on Smart Structures and Materials*, SPIE Smart Materials and Structures Publication No. SPIE 3987-27, Newport Beach, California, 2000, pp. 187–192.
16. A. Antoni Sánchez-Ferrer and H. Finkelmann, Thermal and Mechanical Properties of New Main-Chain Liquid-Crystalline Elastomers, X Congreso Nacional de Materiales (Donostia - San Sebastián, 18-20 Junio 2008), 2008.
17. P. G. de Gennes, M. Hebert and R. Kant, Artificial Muscles Based on Nematic Gels, *Macromol. Symp.*, 1997, **113**, 39.
18. M. Hebert, R. Kant and P. G. de Gennes, Dynamics and Thermodynamics of Artificial Muscles Based on Nematic Gels, *J. Phys. I*, 1997, **7**, 909–918.
19. H. Finkelmann and M. Shahinpoor, Electrically-Controllable Liquid Crystal Elastomer-Graphite Composites Artificial Muscles, *Proceedings of SPIE 9th Annual International Symposium on Smart Structures and Materials*, SPIE Publication No. 4695-53, San Diego, California, March, 2002.

20. M. Shahinpoor and H. J. Schneider, Overview of Liquid Crystal Elastomers, Magnetic Shape Memory Materials, Fullerenes, Carbon Nanotubes, Non-Ionic Smart Polymers and Electrorheological Fluids As Other Intelligent and Multi-Functional Materials, in *Intelligent Materials*, Royal Society of Chemistry, Cambridge, 2008, pp. 499–513.

21. M. Shahinpoor, Further Comments on the Stress Tensor in Nematic Liquid Crystals, *Rheol. Acta*, 1978, **17**(2), 108–109.

22. M. Shahinpoor, A Reformulation of Ericksen – Leslie Continuum Theory of Nematic Liquid Crystals, *Iran. J. Sci. Technol.*, **6**(2), 1977, pp. 103–105.

23. M. Shahinpoor, On the Stress Tensor in Nematic Liquid Crystals, *Rheol. Acta. J.*, 1976, **15**(2), 99–103.

24. M. Shahinpoor, Effect of Material Nonlinearity on the Acceleration Twist Waves in Liquid Crystals, *Mol. Cryst. Liquid Cryst. J.*, 1976, **37**, 121–126.

25. M. Shahinpoor, Finite Twist Waves in Liquid Crystals, *Q. J. Mech. Appl. Math.*, 1975, **XXVIII**, 244–253.

26. M. Carme Calderer, G. Garzon, A. Carlos and B. Yan, A Landau-de Gennes Theory Of Liquid Crystal Elastomers, *Discrete and Continuous Dynamical Systems Series*, vol. 8, no. 2, 2015, pp. 283–302.

21 Hydrogels, Including Chemoresponsive Gels, as Smart Materials

Hans-Jörg Schneider

FR Organische Chemie, Universität des Saarlandes, D 66041 Saarbrücken, Germany
Email: ch12hs@rz.uni-sb.de

21.1 Introduction

Hydrogels can be used for many applications, and are increasingly also being developed in view of their possible biocompatibility.[1–7] Such smart materials can, depending on suitable chemical components, bind or release, for example, drugs,[8] pollutants,[9] catalysts, *etc.*, upon interaction with external effectors, and swell or shrink under the influence of different pH, various chemical compounds,[10] temperature, or light. A particulary often-used technique relies on sol–gel phase transition, where a drug bound to the gel can be released *via* transition to the liquid (sol) state.[10b,c] Most hydrogels are amorphous, some are semicrystalline mixtures of amorphous and crystalline phases, or are crystalline. Hydrogels have a water content typically between 80 and 99%, which can be change by external stimuli; this is the basis of many applications. Natural sources of hydrogels are, for example, agarose, chitosan, methylcellulose or hyaluronic acid, but most smart hydrogels are based on synthetic polymers or rely on chemical modification of natural systems.

Fundamentals of Smart Materials
Edited by Mohsen Shahinpoor
© The Royal Society of Chemistry 2020
Published by the Royal Society of Chemistry, www.rsc.org

Synthetic polymers for gels are usually obtained by copolymerization or cross-linking free-radical polymerization, where hydrophilic monomers are reacted with multifunctional cross-linkers. One can produce polymer chains *via* chemical reaction, by photochemical processes, or *via* radiation for the generation of free radicals. Alternatively, one can modify existing polymers by chemical reaction.

Figure 21.1 illustrates the formation of a hydrogel from a lysine-containing gelator assembled into short fibrils at less than 1% critical gelation concentration (CGC), where ultrasound accelerates the gelation with the formation of nanofibers with average diameters of between 30–40 nm.

Gelators can be neutral or cationic, often in the form of protonated polyamines, or anionic, often in the form of polyacids, or zwitterions, with both positive and negative charges.[11,12] Many efficient gelators are amphiphilic, based on longer more hydrophobic chains equipped with polar groups. Relatively small molecules can also act as quite effective gelators. The examples in Scheme 21.1 show for some hydrogelators that the relationship between structure and efficiency, which is measured as the critical gelation concentration (CGC, in wt.%), is far from straightforward.

Swelling occurs in the first step *via* hydration of the polar, most often ionic groups, then by further uptake of so-called secondary bound water between the hydrophobic parts of the network. Additional swelling is opposed by the covalent or physical cross-links, leading to an elastic network retraction force. Thus, the hydrogel will reach an equilibrium swelling level. Due to the osmotic driving force of the network chains towards infinite dilution, additional absorbed or "free" water can fill larger pores or voids, until it is limited by the cross-linking parts. The swelling ratio can be determined by comparing the weight of the gel portion in the swollen state and the weight in the dry state, or by measuring it *via* microscopy. The ratio of free to bound water in hydrogels can be evaluated by differential scanning calorimetry (DSC), based on the assumption that only free water will be frozen and show up in the endotherm measured. In ^{1}H NMR spectra, free and bound water can be distinguished by different signals.

Double network (DN) hydrogels have recently been introduced as materials that show remarkable mechanical strength and toughness.[13] These hydrogels can also serve as selective, and sensitive sensors of analytes, responding to changes in pH, solvent, light, mechanical force, or electric and magnetic field. DN hydrogels are

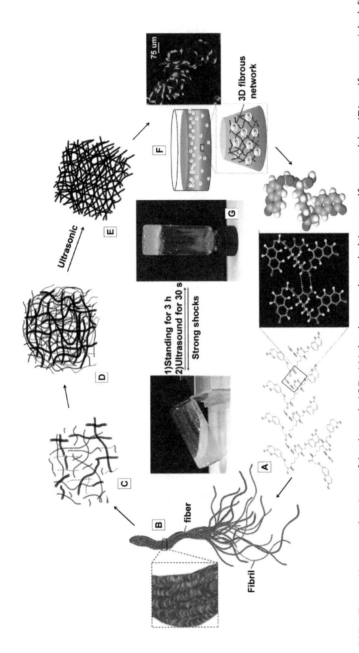

Figure 21.1 Formation mechanism of hydrogel 465: (A) hydrogen bond-driven self-assembly, (B) self-assembled fibrils, (C) fibrils with a hydrogelator concentration of lower than the minimum gelation concentration (MGC), (D) entangled fibrils with a hydrogelator concentration higher than the MGC, (E) well-organized 3D hierarchical nanoarchitectures that form upon ultrasound treatment, (F) cells seeded in hydrogels, (G) optical image of the hydrogel (the transition from solution to hydrogel is reversible). Reproduced from ref. 26 with permission from the Royal Society of Chemistry.

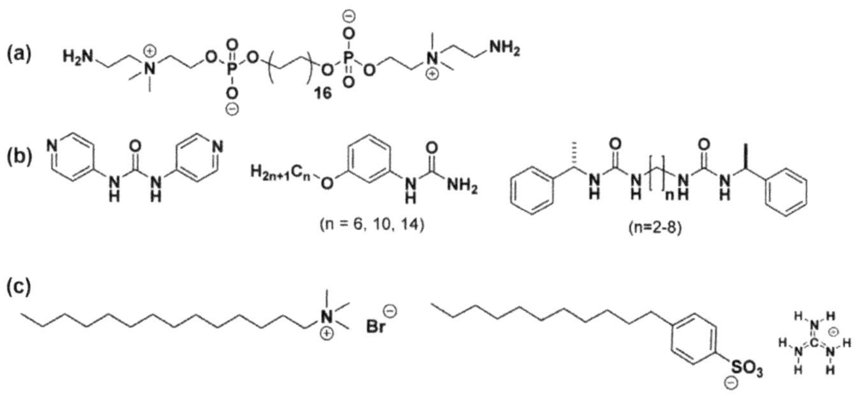

Scheme 21.1 Some representative structures of hydrogelators: (a) critical gelation concentration (CGC) 0.01%, (b) CGC ≈ 1%, and (c) CGC = 4–10%.
Reproduced from ref. 2 with permission from the Royal Society of Chemistry.

usually synthesized *via* a two-step polymerization method, as illustrated in Figure 21.2. In the first step, strong polyelectrolytes form a covalently cross-linked, rigid and brittle gel by UV photopolymerization. The resulting polyelectrolyte gel is immersed and swelled in a precursor solution containing neutral second-network monomers, an initiator, and cross-linker. The second network reactants slowly diffuse into the first gel with ensuing large volume expansion.

In the swelling process, a second polymerization process occurs to form a loosely cross-linked, soft and ductile, second network within the first network. It has been shown that, for example, ultrathin poly(2-acrylamido-2-methylpropanesulfonic acid) (PAMPS)–polyacrylamide (PAAm) DN gels exhibit solvent-triggered force generation within several tens of seconds, acting as fast-responsive artificial muscles.

21.2 Chemoresponsive Materials Based on Hydrogels

Materials for the selective detection of chemical compounds and related actuators can be based on (a) piezoelectric materials (quartz balance), (b) atomic or chemical force microscopy (AFM, CFM) and (c) a variety of other special techniques. The swelling or shrinking of hydrogels is particularly promising, as these processes can be induced

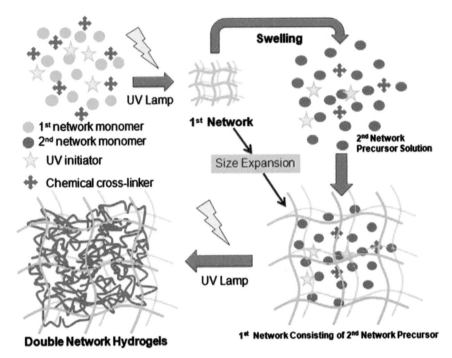

Figure 21.2   The two-step polymerization method used to prepare chemically linked DN hydrogels.
Reproduced from ref. 13 with permission from the Royal Society of Chemistry.

by many external stimuli, *e.g.*, by a pH change in the surrounding liquid.[14] For ionic polymers, this is the consequence of water uptake or release, as the presence of ionic groups is invariably accompanied by solvation water. Furthermore, the uptake of ions increases the osmotic pressure inside the gel due to the formation of the Donnan potential, and thus leads to swelling.

Size changes as a function of pH are illustrated in Figure 21.3 in a hydrogel based on poly(methyl acrylate), which has both cationic and anionic substituents in the form of amino and carboxylic groups. Under neutral conditions, less solvation water is needed, with an ensuing symmetrical profile and a minimum gel particle size at pH 7, also depending on salts in the medium. Such a symmetric profile is typical for gels containing both cationic and anionic sites; the minimum can be shifted *via* the choice of ionic sites with different pK values.

The selectivity of the hydrogel size changes towards different chemical stimuli depends on the binding groups within the

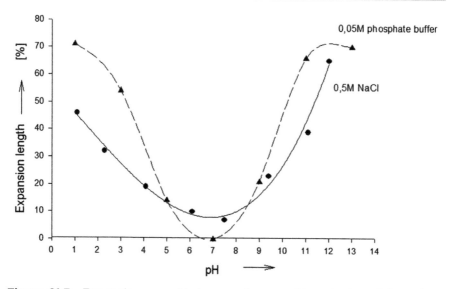

Figure 21.3 Expansion *vs.* pH; in a polymer with a hydrogel based on poly(methyl acrylate), which has both amino and carboxylic groups.
Reproduced from ref. 25 with permission from the American Chemical Society, Copyright 2002.

polymer. Often one observes an increase in swelling in line with the size of the effector molecules, due to taking in of water. Contraction occurs if the effector molecules serve as a non-covalent crosslinking agent.

Significant selectivity can be reached *via* the implantation of supramolecular recognition groups in the polymer backbone, as has been illustrated in a polymerized crystalline colloidal array (PCCA) sensor, which is equipped with crown-ether moieties. Here, micro-capsules of poly *N*-isopropylacrylamide are shown with K ions, a de-creased lower critical solution temperature and swelling or shrinking (Figure 21.4).[15]

Figures 21.5 and 21.6 illustrate how a hydrogel can be used, *e.g.* in microfluidic systems,[16] relying on valves which open or close as a function of the fluid stream pH.[17] By miniaturization, one can reach response times of only 10 seconds and less.

Cyclodextrins (CDs) are frequently used as a non-covalent binding element in supramolecular gels. Figure 21.7 shows an application for a metal-controlled adhesion on the basis of a polyacrylamide hydrogel modified with both β-cyclodextrin (βCD) moieties and 2,2′-bipyridyl moieties.[18] Only in the presence of Zn^{2+} or Cu^{2+} metal ions is the

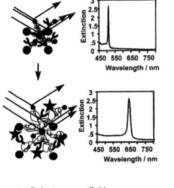

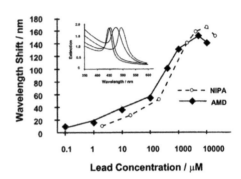

● Polystyrene colloid.

⊃- Side group capable of molecular recognition.

★ Substrate to be recognized.

〰 Hydrogel matrix.

Figure 21.4 (a) Schematic description of a polymerized crystalline colloidal array (PCCA) sensor; the CCA Bragg diffraction monitors the hydrogel volume change induced as a result of the interaction of the polymer side group with a substrate and (b) shift in wavelength as a function of Pb^{2+} concentration with two crown ether-bearing polymers where the inset shows the extinction spectra of one sensor for 0.1, 10, and 100 μM Pb^{2+}.
Reproduced from ref. 15 with permission from the American Chemical Society, Copyright 1998.

bipyridyl residue expelled from the CD cavity, allowing the ßCD unit to interact with another gel chain, which contains *t*-butyl groups as a good binder to the CD group. With, 0.1 M $CuCl_2$ the measured adhesion strength between the particles was as high as 1000 ± 200 Pa, and the addition of strong metal ion chelators such as EDTA was found to reverse the adhesion.

Artificial muscles or actuators, which are triggered selectively by interacting with their chemical environment, with several recognition units have been realized in the form of hydrogels. Figure 21.8 shows the movement of a chitosan hydrogel filled in a small tube, which has pores for contact with the surrounding solution.[19] A 200 g weight could be lifted by a factor of five by swelling induced by acetic acid; in pure water, chitosan contains only about 50% water, in 50 mM acetic acid the water content increases to 96%.[20] The forces are strong due to the water-like low compressibility of the hydrogel, and the containment within the tube also provides a simple, practical solution to the problem of obtaining a unidirectional movement with molecular motors.

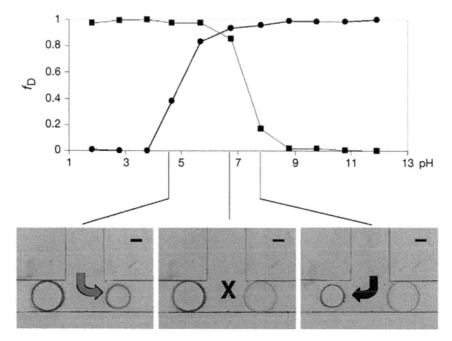

Figure 21.5   Top, the fractional change in diameter (f_D) of the hydrogels with respect to pH. Bottom, a device that directs a fluid stream based on its pH. The hydrogel gating the right branch (circles) expands in base and contracts in acid. The hydrogel gating the left branch (squares) behaves in the opposite manner; at pH 4.7 the flow is directed down the right branch. Both hydrogels expand to shut off the flow when the pH is changed to 6.7.
Reproduced from ref. 17 with permission from Springer Nature, Copyright 2010.

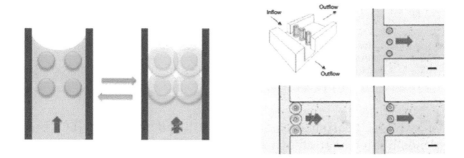

Figure 21.6   Volumetric control of microfluidics with an illustration of a typical set-up.
Reproduced from ref. 17 with permission from Springer Nature, Copyright 2010.

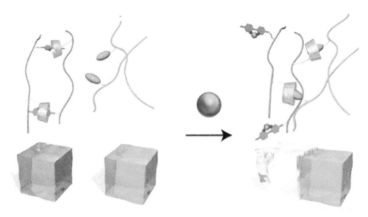

Figure 21.7   A gel particle bearing both β-cyclodextrin (CD) and bipyridyl
moieties reacts with a particle bearing *t*-butyl groups. In the
presence of a metal ion, the CD is set free and associates with
the *t*-butyl gel.
Reproduced from ref. 18, https://doi.org/10.1038/ncomms5622,
under the terms of the CC BY 4.0 licence, https://
creativecommons.org/licenses/by/4.0/.

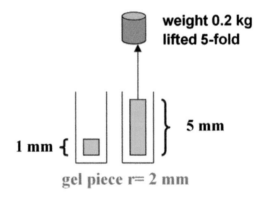

Figure 21.8   The mechanical motion of a chitosan gel *via* expansion trig-
gered by an acid.
Adapted from ref. 19 with permission from the Royal Society of
Chemistry.

A hydrogel bearing cyclodextrin (CD) and ferrocene (FC) units
expands upon oxidation by Ce^{4+}, as the FC^{+} cation no longer binds
to the CD cavity (Figure 21.9).[21] The resulting movement was demonstrated by moving a weight attached to a gel strip and the redox
cycle was found to be completely reversible.

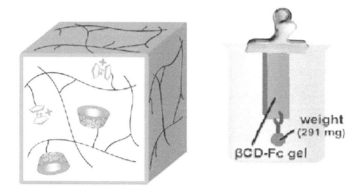

Figure 21.9   A redox-triggered artificial muscle with a gel incorporating both CD and ferrocene FC units.
Adapted from ref. 21 with permission from John Wiley and Sons, Copyright © 2013 WILEY-VCH Verlag GmbH & Co. KGaA, Weinheim.

Hydrogels with chiral units such as those derived from natural chitosan are able to transmit selective motions *via* interaction with different enantiomers.[22] Such chiral discrimination is illustrated as a selective volume variation in Figure 21.10, with enantiomeric dibenzoyl tartaric acids as effectors.[23] The NMR spectra elucidate the interaction mechanism and there is a highly ordered supramolecular complex within the gel, with carboxylate ion pairing between the analyte and the basic polymer backbone, and critical phenyl ring C–H π interactions.

Hydrogels based on peptides, hold great promise in tissue engineering,[23] especially in cell cultures. Peptides that bearing adamantine residues like that in Scheme 21.2 form hydrogels in which phase transition can be achieved by adding β-cyclodextrin (β-CD) derivatives. The disulfide bonds in the gels can be cleaved *via* reduction with glutathione (GSH). The disulfide bond acts as a cleavable linker for control of the molecular self-assembly and thus, the formation of the hydrogel.[24] The hydrogels are formed by disulfide bond reduction using GSH, which is biocompatible for the encapsulation of cells and drugs. It has been shown that mouse fibroblast 3T3 cells attach and grow well at the surface of such hydrogels. Post-culture, 3T3 cells can be recovered from the gels by the addition of a β-CD derivative, which leads to clear solutions as a result of the inclusion of an adamantane residue in the β-CD.

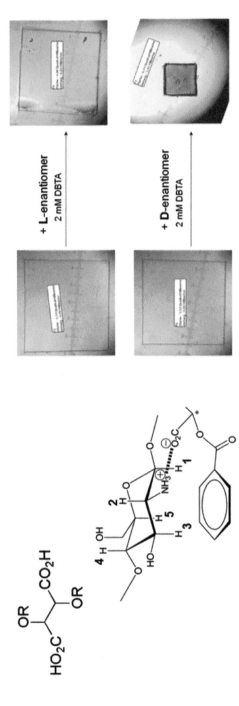

Figure 21.10 A direct translation of chiral recognition to different macroscopic motions, based on the interaction between enantiomeric dibenzoyl tartaric acids (DBTA, R = COPhe) and chitosan gel film pieces.

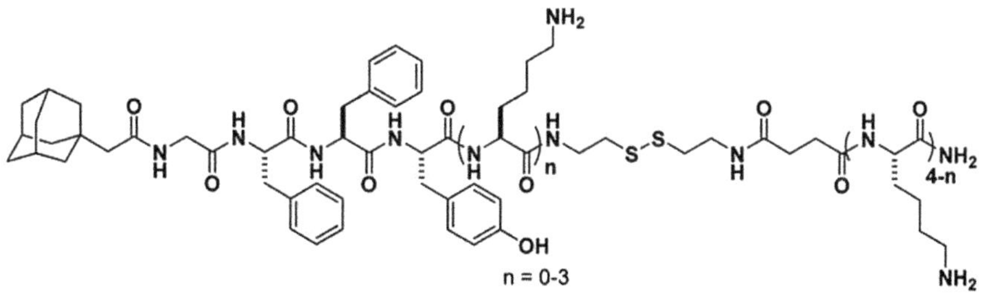

Scheme 21.2 A peptide bearing adamantine residues and disulfide bonds that can be cleaved *via* reduction with glutathione. Sol–gel-transitions can be used in cell culturing.

Homework Problems

Homework Problem 21.1

Propose hydrogel structures with expected swelling or contraction

(a) as a function of pH, symmetric with a minimum at pH 5
(b) as a function of redox agent concentration
(c) in the presence of external effectors such as
    (i) Ca^{2+} ions
    (ii) Cu^{2+} or Zn^{2+} ions
    (iii) glucose
    (iv) phenylalanine

Homework Problem 21.2

Propose hydrogel structures that should exhibit contraction instead swelling with a suitable effector.

Homework Problem 21.3

Design a system for insulin delivery as a function of glucose levels in the blood

(a) as a mechanical pump
(b) as a self-regulating gel particle

References

1. *Polymeric and Self Assembled Hydrogels*, ed. X. J. Loh and O. A. Scherman, Royal Society of Chemistry, Cambridge, 2013.

2. X. Du, J. Zhou, J. Shi and B. Xu, *Chem. Rev.*, 2015, **115**, 13165–13307.
3. L. A. Estroff and A. D. Hamilton, *Chem. Rev.*, 2004, **104**, 1201–1217.
4. N. M. Sangeetha and U. Maitra, *Chem. Soc. Rev.*, 2005, **34**, 821–836.
5. J. A. Foster and J. W. Steed, Exploiting Cavities in Supramolecular Gels, *Angew. Chem., Int. Ed.*, 2010, **49**, 6718–6724.
6. S. S. Babu, V. K. Praveen and A. Ajayaghosh, *Chem. Rev.*, 2014, **114**, 1973–2129.
7. *Nanogels for Biomedical Applications*, ed. A. Vashist, A. K. Kaushik, S. Ahmad, J. Millia Islamia and M. Nair, Royal Society of Chemistry, Cambridge, UK, 2017.
8. *Smart Materials for Drug Delivery*, ed. C. Alvarez-Lorenzo and A. Concheiro, Royal Society of Chemistry, Cambridge, UK, vol. 1,2, 2013.
9. *Smart Materials for Advanced Environmental Applications*, ed. P. Wang, Royal Society of Chemistry, Cambridge UK, 2016.
10. (a) *Chemoresponsive Material*, ed. H.-J. Schneider, RSC, Cambridge, UK, 2015; (b) ch. 2, 5, 6, 9 and 12 in ref. 10a; (c) J. F. Miravet and B. Escuder, Supramolecular Gels for Pharmaceutical and Biomedical Applications 331, in *Supramolecular Systems in Biomedical Fields*, ed. H.-J. Schneider, Royal Society of Chemistry, Cambridge, 2013; (d) see chapter 15 of this book.
11. A. Brizard, R. Oda and I. Huc, in *Low Molecular Mass Gelators: Design, Self-assembly, Function*, ed. F. Fages, *Topics in Current Chemistry*, Springer, Berlin, vol. 256, 2005.
12. P. Dastidar, *Chem. Soc. Rev.*, 2008, **37**, 2699–2715.
13. Q. Chen, H. Chen, L. Zhu and J. Zheng, *J. Mater. Chem. B*, 2015, **3**, 3654–3676.
14. First report: T. Tanaka, D. Fillmore, S.-T. Sun, I. Nishio, G. Swislow and A. Shah, *Phys. Rev. Lett.*, 1980, 1636–1639.
15. (a) J. H. Holtz, J. S. W. Holtz, C. H. Munro and S. A. Asher, *Anal. Chem.*, 1998, **70**, 78; see also, (b) S. W. Pi, X. J. Ju, H. G. Wu, R. Xie and L. Y. Chu, *J. Colloid Interface Sci.*, 2010, **349**, 512, and references cited therein.
16. D. H. Kang, S. M. Kim, B. Lee, H. Yoon and K.-J. Suh, *Analyst*, 2013, **138**, 6230.
17. D. J. Beebe, J. S. Moore, J. M. Bauer, Q. Yu, R. H. Liu, C. Devadoss and B.-H. Jo, *Nature*, 2000, **404**, 588.
18. T. Nakamura, Y. Takashima, A. Hashidzume, H. Yamaguchi and A. Harada, *Nat. Commun.*, 2014, **5**, 4622.
19. K. Kato and H.-J. Schneider, *J. Mat. Chem.*, 2009, **19**, 569.
20. H.-J. Schneider and K. Kato, *Angew. Chem., Int. Ed.*, 2007, **46**, 2694.
21. M. Nakahata, Y. Takashima, A. Hashidzume and A. Harada, *Angew. Chem., Int. Ed.*, 2013, **52**, 5731.
22. D. K. Smith, *Chem. Soc. Rev.*, 2009, **38**, 684.
23. *Smart Materials for Tissue Engineering*, ed. Q. Wang, Royal Society of Chemistry, Cambridge, vol. 1,2, 2017.
24. C. H. Yang, D. X. Li, Z. Liu, G. Hong, J. Zhang, D. L. Kong and Z. M. Yang, *J. Phys. Chem. B*, 2012, **116**, 633–638.
25. H. J. Schneider, T. J. Liu and N. Lomadze, *Eur. J. Org. Chem.*, 2006, 677–692.
26. S. Pan, S. Luo, S. Li, Y. Lai, Y. Geng, B. He and Zhongwei Gu, *Chem. Commun.*, 2013, **49**, 8045–8047.

22 Smart Nanogels for Biomedical Applications

Arti Vashist,[*a] Ajeet Kaushik,[a,b] Srinivasan Chinnapaiyan,[a] Atul Vashist[c] and Madhavan Nair[*a]

[a] Center of Personalized Nanomedicine, Institute of Neuroimmune Pharmacology, Department of Immunology & Nanomedicine, Herbert Wertheim College of Medicine, Florida International University, Miami, FL-33199 USA; [b] Department of Natural Sciences, Division of Sciences, Art, & Mathematics, Florida Polytechnic University, Lakeland, Florida 33805, USA; [c] Department of Biotechnology, All India Institute of Medical Sciences, New Delhi 110029, India
*Emails: avashist@fiu.edu; nairm@fiu.edu

22.1 Introduction

22.1.1 Towards a Hydrogel to Microgel to Nanogel

In recent years, hydrogels have emerged as upcoming biomaterials for use in pharmaceutical science and various biomedical applications. Their journey, which began in 1960, has emerged at great pace. The 21st century presents a huge market for hydrogels with various therapeutics and diagnostics applications.[1] To begin with, hydrogels are defined as the three dimensional (3D) interpenetrating networks of hydrophilic polymers, which have great tendency to absorb a large amount of water and are capable of modifying their characteristic features according to a change in external stimuli.[2] These hydrogels are synthesized using different polymerizations techniques, cross-linking reactions, and click chemistry. They exist in different forms, as thin films,[3] patterned hydrogels,[4] injectable hydrogels,[5] micro

Fundamentals of Smart Materials
Edited by Mohsen Shahinpoor
© The Royal Society of Chemistry 2020
Published by the Royal Society of Chemistry, www.rsc.org

hydrogels,[6] and nano hydrogels. In the present chapter, we are going to have a brief discussion on the unique properties and applications of micro and nano hydrogels.

Micro/nanogels can be defined as gel particles of any shape with a diameter of a micrometer or 1–100 nanometers. These gels are developed using synthetic and natural polymers *via* various polymerization techniques. Although the invention of nanogels was reported long ago by Sager and coworkers in 1989, since then efforts have been made in the development of nanogels for a diverse range of biomedical applications. Recent trends show that click chemistry has been employed for the development of various forms of hydrogels, including injectable, patterned hydrogels and also in micro and nanogel forms, which justify the diversity of applications exhibited by hydrogels.

The ongoing research shows great progress in the applications of both microgel and nanogel particles for numerous applications, including chemotherapy,[7] sensing,[8] cell imaging,[9] targeting, tissue engineering[10] and 3D printing technology. A responsive nanogel can be designed to achieve targeted drug delivery[11] (Figure 22.1).

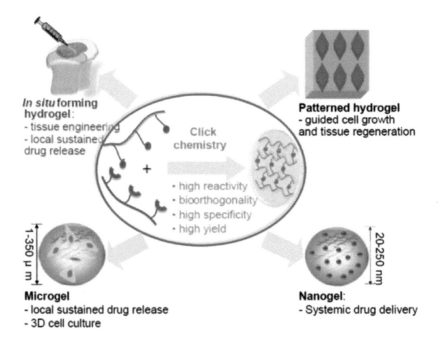

Figure 22.1 Preparation and potential biomedical applications of click hydrogels, microgels, and nanogels.
Reproduced from ref. 12 with permission from Elsevier, Copyright 2012.

The hydrophilic nature of nanogels results in their high bio-compatibility and high loading capacity. The integrity of the drug is protected inside the nanogel, and its release can be actively controlled by the modulation of the structure and functionality of the nanogel.

Nanogels hold great potential to withstand degradation from enzymes and are thus being utilized extensively for nucleic acid (RNA, DNA) encapsulation. The high stability of the nanogel system helps in achieving increased half-lives for small molecules and thus offers a unique platform for the combined delivery of therapeutics. Targeted delivery is achieved by tagging with antibodies or specific targeting ligands. The aim of the present chapter is to give a brief introduction of the synthetic methodology of micro/nanosystems, their characteristic features, the polymers and nanofillers employed in their synthesis, and their various biomedical applications.

22.2 Polymer-based Micro/Nano Gels

Biopolymers and synthetic polymers are being deployed in the development of nanogels. The use of biodegradable and biocompatible polymers in nanomedicine applications makes nanogels more successful for therapeutic applications. Table 22.1 summarizes the important and common biopolymers as well as synthetic polymers used for the synthesis of nanogels. Chitosan with glucosamine units that have reactive amino and hydroxyl groups has been shown to actively bind various drugs and have bioactivity, and is thus being used for developing responsive chitosan-based nanogels. Other biopolymers include sodium alginate, which is also commonly used for nanogel synthesis because it is one of the components of the biofilm produced by the bacterium *Pseudomonas Aeruginosa*, which is a well-known

Table 22.1 Various natural and synthetic polymers used in the development of nanogels.

Natural polymers	Ref.	Synthetic polymers	Ref.
Chitosan and chitin	13	Polyethylene glycol	14
Gelatin	15	Polyvinylalcohol	16
Polypeptides	17	Poly(*N*-isopropylacrylamide)	18
Sodium alginate	19	Poly(lactic acid)	20
Hyaluronic acid	21	Poly(glycolic acid)	22
Dextran, carboxydextran	23	Poly(ethyleneimine)	24
Elastin	25	Polyacrylic acid	26
Cellulose and its derivatives	27	Pluronics	28
		Polyamine	29

pathogen in cystic fibrosis, and hence is very resistant to antibiotics as well as macrophages.

22.3 Synthesis of Micro/Nanogels

The synthesis of gel particles in micro as well as nanoforms is carried out using various emulsion polymerization techniques or copolymerization.[30] Emulsion techniques use water-in-oil or oil-in-water polymerization process. The size of particles is controlled by the concentration of the surfactant and the crosslinker. Stirring time and rate are also important parameters to use to alter the formation of homogeneous particles. The other techniques employed for synthesis include precipitation, solvent diffusion, and spray drying. Molecular imprinting (MIT) and the milling process are also emerging techniques that can be used to develop nanogel particles. Bottom-up and top-down approaches are commonly known methodologies used for nanogel synthesis. Nanogel synthesis includes various chemical interactions and hydrophobic modification, which results in the development of stable nanogel particles. Self-assembly results in noncovalent interactions. The various charges present on the polymers used for the synthesis are also a key factor in developing a specific charge carrying nanogel, which is the utmost requirement for tagging and binding various bioactive compounds inside or within the nanogel. The other method employed for nanogel development is the membrane emulsification method, which includes a dispersed phase that can be filtered by passing it through a ceramic membrane with pressure applied, thus resulting in droplets of the dispersed phase with uniform particles forming on the membrane. Heterogeneous free radical polymerization techniques are also used for nanogel preparation.

22.4 Characterization of Nanogels

This section will introduce the basic characterization techniques that should be carried out in order to fully determine the structural and physicochemical properties of nanogels. The first step to understand the formation of the interpenetrating networks is to perform functional group characterization by the type of crosslinking that occurs within the nanogel network, which is commonly achieved using Raman and Fourier-transform infrared (FTIR) spectroscopy. Nuclear magnetic resonance (NMR) spectroscopy is also utilized to completely elucidate the structure and geometry of the synthesized nanogel particles. These studies completely reveal the exact crosslinking reaction that occurs during the formation of the interpenetrating network. Mass spectroscopy and gel permeation chromatography are carried out to determine the exact molecular weight of the newly synthesized nanogels, which gives a clear idea about their structure. Differential light scattering (DLS) is performed to know the size and distribution of the particles. Furthermore, the shape and size of the nanogel particles are confirmed by transmission electron microscopy (TEM) analysis. The nanogels are characterized by UV–vis spectrophotometry and high-performance liquid chromatography (HPLC) to analyze their drug encapsulation efficiency and percentage of drug release. Also, their crystallinity or amorphous behavior can be determined by X-ray diffraction (XRD). Degradability studies *via* hydrolytic or enzyme degradation are carried out using various buffer solutions of different pH and other enzymes according to the applications in which they are being deployed. Biocompatibility testing is carried out using various *in vitro* cell viability assay methods, for example, XTT and MTT assays kits. These are preliminary characterization techniques that have to be carried out after the development of the nanogels. Further depending upon the applications of the nanogels, they are then tested for specific applications.

22.5 Biomedical Applications

Nanogels are being used for various biomedical applications such as gene therapy, imaging, drug delivery, biosensing, and tissue regeneration. In the next section, we will give a brief overview of a few important studies that have been carried out in recent years using nanogel technology for specific applications in medicine. The most important features of nanogels that have emerged in recent years is that they can act as imaging agents and are being applied in diagnostics.[31] Their structure, water absorbing features, and compatibility with physiological fluids mean that they can be encapsulated with various imaging agents such as quantum dots, dyes, or metal ions. These imaging structures are very stable inside the nanogels, thus making them unique imaging agents.[9] The cytotoxicity and other toxicity levels of inorganic particles are greatly reduced when they are encapsulated inside nanogels. Many research studies that have focused on the use of gold nanoparticles with nanogels have shown promising results in imaging. Nanogels are innovative drug delivery systems, and can be used for numerous biomedical applications for the therapeutic targeting of and drug delivery for the treatment of cancer, diabetes, inflammation, bone regeneration, bacterial infections, viral infections, brain diseases, *etc.* Nanogels have been extensively used in terms of imaging, diagnostics, drug delivery, tissue engineering, and biosensing, and have shown great biocompatibility and biodegradability both in *in vivo* and *in vitro* studies. In addition to the enormous potential of functional nanogels as novel polymeric platforms for biomedicine, drug and gene delivery, and smart imaging, biomedical and pharmaceutical applications of nanogels have been fully explored for tissue regeneration, wound healing, surgical device, implantation, and peroral, rectal, vaginal, ocular and transdermal drug delivery (Figure 22.2). Several studies have also reported that nanogels are able to alter their stability in an environmentally sensitive manner, under the influence of temperature, pH, and glutathione concentration, where a gel has been shown to demonstrate thermally-responsive cellular uptake of payloads through electrostatic absorptive endocytosis. Prominently, this thermally-responsive nanogel carrying paclitaxel exhibited outstanding anticancer efficacy in HT-29 human colon cancer-bearing xenograft mice. Chen and co-workers[32] reported and prepared dual thermo- and pH-sensitive micellar nanogels composed of mPEG–isopropylideneglycerol, which is

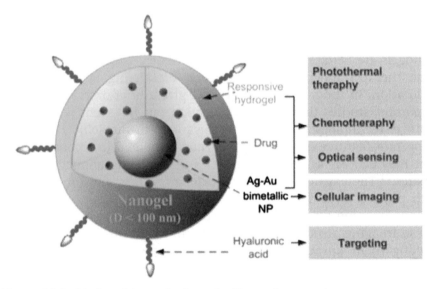

Figure 22.2   Various biomedical applications of nanogels.
Reproduced from ref. 33 with permission from Elsevier, Copyright 2010.

promising for use in the targeted delivery of payloads, and enhancing the adverse effects of cancer.

In another study, Eckmann and colleagues explored biocompatible physically crosslinked hybrid nanogels consisting of partially denatured lysozyme cores and dextran shells for the local delivery of dexamethasone to alleviate acute pulmonary inflammation.[34] The nanogels were coated with antibodies directed to the endothelial determinant to target the pulmonary vasculature, intercellular adhesion molecule-1 (ICAM). The synthesized ICAM-targeted nanogels were loaded with dexamethasone at 5 wt.%. Incomparable to other research, dextran–lysozyme nanogels were utilized as scaffolds for the *in situ* synthesis of silver nanoparticles.[35] Such hybrid nanogels exhibited bacteriostatic properties towards *S. aureus* and bacterial properties towards *E. coli*, and their antibacterial activity was altered by varying the lysozyme content. These gels make it possible to optimize the release of a bactericidal agent for specific clinical use according to the type of infection.[36] PEGylated nanogels containing gold nanoparticles were developed in a core of poly(2-[*N*,*N*-diethylamino]ethyl methacrylate) (PEAMA) for cancer photothermal therapy. No reducing agent was used in this process.[37]

These gold-containing nanogels showed remarkable photothermal efficacy in response to laser-irradiation, resulting in selective cytotoxicity in cancer cells and enhanced cancer cell radiosensitivity.[38]

Nanogels have been used as a potential candidates for the delivery of different oligonucleotides, such as antisense oligodeoxynucleotides, small interfering RNAs, micro RNAs designed for the therapeutic targeted inhibition of specific mRNA sequences that are of emerging interest for the treatment and cancer diagnosis, neurodegenerative disorders and lethal viral infections (Dias and Stein, 2002; Seidman *et al.*, 1999; Janssen *et al.*, 2013; Kanasty *et al.*, 2013).[39–42] Some of the oligonucleotide-based therapies have already achieved significant success in clinical trials. Some other nanogels that have tested as vaccine carriers have also been found to induce significant responses associated with protective immunity against selected bacterial and viral infections.[43] Senanayake *et al.* (2015) demonstrated that nano drugs show more favorable pharmacokinetics compared to free nucleoside reverse transcriptase inhibitors (NRTIs). Infrequent I.V. injections of PEGylated CEPL–sAZT alone was found to significantly suppress HIV-1 RT activity to background levels in a humanized mouse (hu-PBL) HIV model.[44] These studies demonstrate the potential of nanogels for use in a diverse range of biomedical applications.

22.6   Conclusion

In conclusion, it is implicated that the combined efforts of materials scientists, nanotechnologists and biotechnologists have resulted in the development of nanogels that have unique properties that are needed for improved performance in the area of targeted drug delivery, tissue engineering, imaging, diagnostics, *etc.* Considering advanced state-of-the-art nanogel technology, these prospective candidates will certainly show improvements in the targeted technology and customization of novel nanogels needed to develop a friendly nanogel platform for desired applications. There is significant potential for these advanced nanostructures to be translated into use in personalized healthcare.

Homework Problems

Homework Problem 22.1

What property should a hydrogel exhibit to have image-guided biomedical applications?

Homework Problem 22.2

What are the characteristic features of hydrogels which make them perfect candidates to be used in drug delivery?

Homework Problem 22.3

What are injectable hydrogels?

Homework Problem 22.4

Which functional groups are responsible for the swelling properties of hydrogels?

Homework Problem 22.5

List a few crosslinkers that can be used for crosslinking of hydrogels.

Homework Problem 22.6

List a few hydrogels that are on the market.

Acknowledgements

The authors acknowledge financial support from the National Institute of Health (NIH) grant RO1DA042706.

References

1. A. Vashist, A. Kaushik, A. Ghosal, R. Nikkhah-Moshaie, A. Vashist, R. Dev Jayant and M. Nair, *Nanogels for Biomedical Applications*, Royal Society of Chemistry, DOI: 10.1039/9781788010481-00001, 2018, pp. 1–8.
2. A. Vashist, A. Vashist, Y. Gupta and S. Ahmad, *J. Mater. Chem. B*, 2014, **2**, 147–166.
3. A. Vashist, Y. K. Gupta and S. Ahmad, *Carbohydr. Polym.*, 2012, **87**, 1433–1439.
4. A. Vashist, S. Shahabuddin, Y. K. Gupta and S. Ahmad, *J. Mater. Chem. B*, 2013, **1**, 168–178.
5. G. Lokhande, J. K. Carrow, T. Thakur, J. R. Xavier, M. Parani, K. J. Bayless and A. K. Gaharwar, *Acta Biomater.*, 2018, **70**, 35–47.
6. M. Hirayama, K. Tsuruta, A. Kawamura, M. Ohara, K. Shoji, R. Kawano and T. Miyata, *J. Micromech. Microeng.*, 2018, 034002.
7. H. Akhter and S. Amin, *Curr. Drug Delivery*, 2017, **14**, 597–612.
8. P. Manickam, M. Pierre, R. Dev Jayant, M. Nair and S. Bhansali, *Nanogels for Biomedical Applications*, Royal Society of Chemistry, DOI: 10.1039/9781788010481-00261, 2018, pp. 261–282.
9. M. Chan and A. Almutairi, *Mater. Horiz.*, 2016, **3**, 21–40.
10. J. Thompson and R. Dua, *Nanogels Biomed. Appl.*, 2017, **30**, 77.
11. A. Vashist, A. Kaushik, A. Vashist, J. Bala, R. Nikkhah-Moshaie, V. Sagar and M. Nair, *Drug Discovery Today*, 2018, **23**(7), 1436–1443.
12. Y. Jiang, J. Chen, C. Deng, E. J. Suuronen and Z. Zhong, *Biomaterials*, 2014, **35**, 4969–4985.
13. A. Masotti and G. Ortaggi, *Mini-Rev. Med. Chem.*, 2009, **9**, 463–469.
14. M. J. Simpson, B. Corbett, A. Arezina and T. Hoare, *Ind. Eng. Chem. Res.*, 2018, 7495–7506.
15. P. R. Sarika, P. R. Anil Kumar, D. K. Raj and N. R. James, *Carbohydr. Polym.*, 2015, **119**, 118–125.
16. H. Moshe, Y. Davizon, M. M. Raskin and A. Sosnik, *Biomater. Sci.*, 2017, **5**, 2295–2309.
17. S. S. Desale, S. M. Raja, J. O. Kim, B. Mohapatra, K. S. Soni, H. Luan, S. H. Williams, T. A. Bielecki, D. Feng and M. Storck, *J. Controlled Release*, 2015, **208**, 59–66.
18. H. R. Culver, I. Sharma, M. E. Wechsler, E. V. Anslyn and N. A. Peppas, *Analyst*, 2017, **142**, 3183–3193.
19. M. Pei, X. Jia, X. Zhao, J. Li and P. Liu, *Carbohydr. Polym.*, 2018, **183**, 131–139.
20. L. N. Woodard, K. T. Kmetz, A. A. Roth, V. M. Page and M. A. Grunlan, *Biomacromolecules*, 2017, **18**, 4075–4083.
21. X. Wei, T. H. Senanayake, G. Warren and S. V. Vinogradov, *Bioconjugate Chem.*, 2013, **24**, 658–668.
22. F. Chai, L. Sun, X. He, J. Li, Y. Liu, F. Xiong, L. Ge, T. J. Webster and C. Zheng, *Int. J. Nanomed.*, 2017, **12**, 1791.
23. C. Gonçalves, J. P. Silva, I. F. Antunes, M. Ferreira, J. A. Martins, C. F. Geraldes, Y. Lalatonne, L. Motte, E. de Vries and F. M. Gama, *Bioconjugate Chem.*, 2015, **26**, 699–706.

24. N. Kordalivand, D. Li, N. Beztsinna, J. S. Torano, E. Mastrobattista, C. F. van Nostrum, W. E. Hennink and T. Vermonden, *Chem. Eng. J.*, 2018, **340**, 32–41.
25. I. González de Torre, L. Quintanilla, G. Pinedo-Martín, M. Alonso and J. C. Rodríguez-Cabello, *ACS Appl. Mater. Interfaces*, 2014, **6**, 14509–14515.
26. R. Dharela, S. Kumari, G. S. Chauhan, J. Manuel and J.-H. Ahn, *Sci. Adv. Mater.*, 2017, **9**, 1280–1284.
27. K. Rahimian, Y. Wen and J. K. Oh, *Polymer*, 2015, **72**, 387–394.
28. R. Ghaffari, N. Eslahi, E. Tamjid and A. Simchi, *ACS Appl. Mater. Interfaces*, 2018, **10**(23), 19336–19346.
29. M. R. Gordon, B. Zhao, F. Anson, A. Fernandez, K. Singh, C. Homyak, M. Canakci, R. W. Vachet and S. Thayumanavan, *Biomacromolecules*, 2018, **19**, 860–871.
30. A. Ghosal, S. Tiwari, A. Mishra, A. Vashist, N. K. Rawat, S. Ahmad and J. Bhattacharya, *Nanogels for Biomedical Applications*, Royal Society of Chemistry, DOI: 10.1039/9781788010481-00009, 2018, pp. 9–28.
31. A. E. Ekkelenkamp, M. R. Elzes, J. F. Engbersen and J. M. Paulusse, *J. Mater. Chem. B*, 2018, **6**, 210–235.
32. D. Chen, H. Yu, K. Sun, W. Liu and H. Wang, *Drug Delivery*, 2014, **21**, 258–264.
33. W. Wu, J. Shen, P. Banerjee and S. Zhou, *Biomaterials*, 2010, **31**, 7555–7566.
34. M. C. C. Ferrer, V. V. Shuvaev, B. J. Zern, R. J. Composto, V. R. Muzykantov and D. M. Eckmann, *PLoS One*, 2014, **9**, e102329.
35. M. C. C. Ferrer, R. C. Ferrier, D. M. Eckmann and R. J. Composto, *J. Nanopart. Res.*, 2013, **15**, 1323.
36. M. C. C. Ferrer, S. Dastgheyb, N. J. Hickok, D. M. Eckmann and R. J. Composto, *Acta Biomater.*, 2014, **10**, 2105–2111.
37. T. Nakamura, A. Tamura, H. Murotani, M. Oishi, Y. Jinji, K. Matsuishi and Y. Nagasaki, *Nanoscale*, 2010, **2**, 739–746.
38. H. Yasui, R. Takeuchi, M. Nagane, S. Meike, Y. Nakamura, T. Yamamori, Y. Ikenaka, Y. Kon, H. Murotani and M. Oishi, *Cancer Lett.*, 2014, **347**, 151–158.
39. N. Dias and C. Stein, *Eur. J. Pharm. Biopharm.*, 2002, **54**, 263–269.
40. S. Seidman, F. Eckstein, M. Grifman and H. Soreq, *Antisense Nucleic Acid Drug Dev.*, 1999, **9**, 333–340.
41. H. L. Janssen, H. W. Reesink, E. J. Lawitz, S. Zeuzem, M. Rodriguez-Torres, K. Patel, A. J. van der Meer, A. K. Patrick, A. Chen and Y. Zhou, *N. Engl. J. Med.*, 2013, **368**, 1685–1694.
42. R. Kanasty, J. R. Dorkin, A. Vegas and D. Anderson, *Nat. Mater.*, 2013, **12**, 967.
43. T. Nochi, Y. Yuki, H. Takahashi, S.-I. Sawada, M. Mejima, T. Kohda, N. Harada, I. G. Kong, A. Sato and N. Kataoka, *Nat. Mater.*, 2010, **9**, 572.
44. T. Senanayake, S. Gorantla, E. Makarov, Y. Lu, G. Warren and S. Vinogradov, *Mol. Pharmaceutics*, 2015, **12**, 4226–4236.

23 Review on Self-healing Materials

Mohsen Shahinpoor

Mechanical Engineering Dept., University of Maine, USA
Email: shah@maine.edu

23.1 Introduction

Briefly introduced and discussed in this chapter is a family of self-healing materials. The self-healing characteristics of these materials, and in particular biomaterials, and the concepts of the self-healing processes in nature and biology are already well known by the scientific communities.[1] One can start by describing their impact and occurrence in nature, in plants, in animals and human beings. These understandings of self-healing processes in biology and nature are particularly more advanced in terms of dermatology and skin repair by scar tissues[2] and they have further led to their most recent industrial applications and scientific discoveries. This chapter will introduce, briefly describe, and explain a wide range of self-healing smart materials. These materials have internal structural abilities and characteristics that enable them to automatically self-repair damage with almost no external intervention or diagnosis. It is well recognized that using various materials over time will degrade them due to several phenomena such as fatigue failure, environmental degradation, or damages such as cracks, fracture, and creep incurred during operation. In general, internal cracks are difficult to detect, and manual intervention may be necessary. The advantage of self-healing

Fundamentals of Smart Materials
Edited by Mohsen Shahinpoor
© The Royal Society of Chemistry 2020
Published by the Royal Society of Chemistry, www.rsc.org

materials is that they can treat material degradation by initiating a repair mechanism that responds to the incurred damage or degradation. Smart materials and structures also play important roles in self-healing materials because they are multi-functional and are capable of handling various environmental conditions.

23.2   Self-healing Materials

The pyramids at Giza in Egypt have survived for thousands of years because the ancient Egyptians learned to use limestone ($CaCO_3$ plus CaO) and its powdered form as a strengthening and healing material, mixed with mortars. The materials science involved has been studied,[3] noting that many of the principles involved in self-healing originate from biological and natural system self-healing mechanisms.[4] Self-healing materials have been recognized as a field of study in engineering in the 21st century. Self-healing materials are capable using external stimuli such as light, motion, sound, humidity, electromagnetic fields, thermal fields, and so on to initiate the self-healing process.[5] In the next sections of this chapter, self-healing concrete, as well as polymeric materials and hybrid composites, are further elaborated on.

23.3   Self-healing Cementitious and Concrete Materials

The reaction of certain chemicals may be used to initiate the self-healing of cementitious materials. Typically, tiny capsules and vascular tubes are embedded within the material such that upon initiation of any micro crack these capsules and tubes can be ruptured and healing agents can be released inside the material to start healing the cracks. Research has focused on improving the quality of these embedded capsules and vascular tubes. Another mechanism of self-healing based on bacterial intelligence has also been implemented. Bacteria embedded within cementitious materials can generate, through their metabolic activities, certain calcium carbonate precipitates or sap to repair any damage. Concrete is quite alkaline, and the "healing" bacteria (Bacillus) thrives in the alkaline environment of the concrete for years in a dormant manner. If there are any cracks in the concrete, then water from outside can get in and activate the dormant Bacillus bacteria, which then secrete calcium carbonate. The Bacillus bacteria thrive in the material and produce spores that can survive for years without food or oxygen, and they remain intact during mixing, only dissolving and becoming active if the concrete cracks and water gets in. The bacteria secrete healing material or limestone ($CaCO_3$) into the concrete. Note that limestone is a sedimentary rock and mainly covered with skeletal fragments of marine organisms, such as coral reefs and mollusks (Figure 23.1).[6,7]

Only just recently in the 21st century, the subject of self-healing materials has emerged and gained importance. The first international conference on self-healing materials was held in 2007.[8] The field of self-healing materials is related to biomimetic materials,[9] Murray's law,[10–16] as well as other novel materials and surfaces with the inherent capacity of self-organization, such as self-lubricating and self-cleaning materials.[9]

Murray's law states that in a branched vasculature of the mammalian circulatory and respiratory system filled with a Newtonian fluid such as water, as well as the xylem in plants with the basic function of trans-porting water and nutrients from roots to branches, shoots and leaves, the cube of the radius of a parent vessel should be equal to the sum of

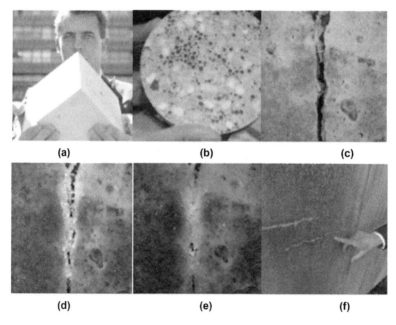

(a) (b) (c)

(d) (e) (f)

Figure 23.1 Embedded self-activating bacteria in concrete that make it self-healing.[31] (a) Shows a block of self-healing concrete, (b) is a picture of the surface, (c) shows a crack, which then starts to heal, (d) and further heals, (e) and is finally fully healed (f). Reproduced from https://tudelft.openresearch.net/page/12638/bio-concrete-a-self-healing-construction-material with permission from the University of Delft.

the cubes of the radii of the daughter branches and shoots. This law also means that a functional relationship exists between vessel radius and volumetric flow rate, the average linear velocity of flow, velocity profile, vessel-wall shear stress, the Reynolds number, and pressure gradient in individual vessels. Thus, the volumetric flow rate is proportional to the cube of the radius in a cylindrical channel optimized for the minimum work to drive and maintain the fluid. By considering the limits of the generalized Murray's law, the optimum daughter–parent area ratio Γ, for symmetric branching into N daughter channels of any constant cross-sectional shape, is $\Gamma^{-2/3}$ for large scale channels and $\Gamma^{-4/5}$ for channels with a characteristic length much smaller than the slip length.[10–16]

23.4   Self-healing Polymers and Elastomers

Polymers have become fundamental materials in our everyday lives and nowadays, products such as plastics, rubbers, films, fibers, membranes or paints, *etc.*, are very common. Thus, we are forced to extend the reliability and maximum lifetime of polymers. A newly designed class of polymeric materials that are capable of self-healing and restoring their functionality after damage or fatigue have been identified.[17,18] Numerous methodologies for the assessment of self-healing capabilities have been developed for each materials class. There is an excellent presentation on self-healing materials by Yang,[18] which lists various materials with their damage mechanisms and their healing processes and mechanisms: (https://en.wikipedia.org/wiki/Self-healing_material#cite_note-Yang-29).

   Note that the self-healing process is a three-step process similar to the self-healing process in biological or botanical systems,[19,20] where the first response is triggering or actuation mechanism, which happens after damage occurs. The second response is the initiation of the transport of materials to the damaged regions, following the actuation. The third response is the chemical repair process, which differs depending on the type of healing mechanism that is in connected to the actual chemical self-healing process. For additional references on self-healing polymeric materials, refer to the relevant literature.[21–27]

23.5   EMAAs as Ionic Self-healing Polymers

Poly(ethylene-*co*-methacrylic acid), or EMAA, is a commercial thermoplastic carboxylic ionomer (a random semi-crystalline polymer in both acidic and neutralized forms), which was introduced by DuPont in the early 1960s, and is commercially trademarked and available as Surlyn®. As discussed in ref. 28–31, Surlyn® is a smart self-healing material that ionically repairs any microcracks within a material by absorbing any impact energy ionically. Surlyn® is a copolymer of ethylene and methacrylic acid groups and is partially neutralized with Na^+ ions. Surlyn® is heavily used in impact protection applications, such as high impact sports like golf, in a golf ball, or in boats or car bumpers. Surlyn® has been shown to have outstanding resistance to both chemical and physical attack.

The inclusion of a few mol% of ionic groups along the backbone has a tremendous effect on the morphology and properties of Surlyn® (Fallahi, Shahinpoor, *et al.*, 2017[31]). Thus, active multi-functional self-healing Surlyn® smart materials can be manufactured. As reported in ref. 31, novel soft robotic materials were developed that are capable of actuation and sensing, and the methods of manufacturing and testing them as multi-functional electroactive polymer–metal nanocomposites is called (EMAMC), based on the REDOX chemical plating of poly-(ethylene-*co*-methacrylic acid) or (EMAA) ionomer with a metal.

The presence of the methacrylic acid units and the neutralized carboxylate anion/cation pairs provides sites for ionic interaction. The EMAMCs show the flexibility of traditional ionic polymers and are thus considered to be strong candidates for soft robotic actuators and sensors, which can be used as soft elbow, finger joint, knee, hip or segmented limb soft actuators, energy harvesters and sensors.[31,32] The advantage of Surlyn®, however, is that it has self-healing properties.

23.6 Conclusions

This chapter presented a family of self-healing materials capable of self-detecting any damage that can kick off self-healing and self-repair processes to overcome the damage. It was concluded that one can start by describing their impact and occurrence in nature, in plants, in animals and human beings. These understandings of self-healing processes in biology and nature are particularly more advanced in terms of dermatology and skin repair by scar tissues, and they have further led to the most recent industrial applications and scientific discoveries. This chapter introduced self-healing materials very briefly and explained a wide range of self-healing smart materials. These materials will have internal structural abilities and characteristics that enable them to automatically self-repair damage with almost no external intervention or diagnosis. It was concluded that there are many advantages of using self-healing materials in the sense that they can treat material degradation by initiating a repair mechanism that responds to the incurred damage. Smart materials and structures also play important roles as self-healing materials because they are multi-functional and are capable of handling various environmental conditions.

Homework Problems

Homework Problem 23.1

Discuss Murray's law in connection with self-healing materials.

Homework Problem 23.2

What are golf balls made up of and what self-healing characteristics do they possess?

References

1. S. K. Ghosh, *Self-healing Materials: Fundamentals, Design Strategies, and Applications*, Wiley Publishers-VCH, Weinheim, 2008, p. 145.
2. C. H. F. Hämmerle and W. V. Giannobile, Biology of soft tissue wound healing and regeneration – Consensus Report of Group 1 of the 10th European Workshop on Periodontology, *J. Clin. Periodontol.*, 2014, **41**, S1–S5.
3. P. Thirumalini, R. Ravi, S. K. Sekar and M. Nambirajan, Study on the Performance Enhancement of Lime Mortar Used in Ancient Temples and Monuments in India, *Indian J. Sci. Technol.*, 2011, **4**(11), 1484.
4. E. Wayman, The Secrets of Ancient Rome's Buildings, Smithsonian Institution, Retrieved 13 November 2016.
5. C. E. Diesendruck, N. R. Sottos, J. S. Moore and S. R. White, Biomimetic Self-Healing, *Angew. Chem., Int. Ed.*, 2015, **54**(36), 10428–10447.
6. H. Jonkers, A. Thijssen, G. Muyzer, O. Copuroglu and E. Schlangen, Application of bacteria as self-healing agent for the development of sustainable concrete, *Ecol. Eng.*, 2010, **36**(2), 230–235.
7. H. Jonkers, Bacteria-based self-healing concrete, *Heron*, 2011, **56**(1/2), 1–2.
8. First international conference on self-healing materials, The Delft University of Technology, 12 April 2007.
9. M. Nosonovsky and P. Rohatgi, Biomimetics in Materials Science: Self-healing, self-lubricating, and self-cleaning materials, *Springer Ser. Mater. Sci.*, **152**, Springer, 2011.
10. C. D. Murray, The Physiological Principle of Minimum Work: I. The Vascular System and the Cost of Blood Volume, *Proc. Natl. Acad. Sci. U. S. A.*, 1926, **12**(3), 207–214.
11. X. Zheng, G. Shen, C. Wang, Y. Li, D. Dunphy, T. Hasan, C. J. Brinker and B. L. Su, Bio-inspired Murray materials for mass transfer and activity, *Nat. Commun.*, 2017, **8**, 14921.
12. T. F. Sherman, On connecting large vessels to small. The meaning of Murray's law, *J. Gen. Physiol.*, 1981, **78**(4), 431–453.

13. H. R. Williams, R. S. Trask, P. M. Weaver and I. P. Bond, Minimum mass vascular networks in multifunctional materials, *J. R. Soc., Interface*, 2008, **5**(18), 55–65.
14. K. A. McCulloh, J. S. Sperry and F. R. Adler, Water transport in plants obeys Murray's law, *Nature*, 2003, **421**(6926), 939–942.
15. C. D. Murray, The Physiological Principle of Minimum Work: II. Oxygen Exchange in Capillaries, *Proc. Natl. Acad. Sci. U. S. A.*, 1926, **12**(5), 299–304.
16. D. Stephenson, A. Patronis, D. M. Holland and D. A. Lockerby, Generalizing Murray's law: An optimization principle for fluidic networks of arbitrary shape and scale, *J. Appl. Phys.*, 2015, **118**(17), 174302–174308.
17. M. Q. Zang, Self-healing in polymers and polymer composites. Concepts, realization, and outlook: A review, *Polym. Lett.*, 2008, **2**(4), 238–250.
18. Y. Yang, Wikipedia, Self-Healing Materials, https://en.wikipedia.org/wiki/Self-healing_material#cite_note-Yang-29.
19. T. Speck, R. Mülhaupt and O. Speck, Self-healing in plants as bio-inspiration for self-repairing polymers, in *Self-Healing Polymers*, ed. W. Binder, Wiley-VCH, 2013, pp. 61–89.
20. T. Speck *et al.*, Bio-inspired self-healing materials, in *Materials Design Inspired by Nature: Function through Inner Architecture, Smart Materials*, P. Fratzl, J. W. C. Dunlop and R. Weinkamer, Royal Society of Chemistry, vol. 4, 2013, pp. 359–389.
21. A. C. Schüssele, F. Nübling, Y. Thomann, O. Carstensen, G. Bauer, T. Speck and R. Mülhaupt, Self-healing rubbers based on NBR blends with hyperbranched polyethyleneimines, *Macromol. Mater. Eng.*, 2012, **9**(5), 411–419.
22. Y. Yang and M. W. Urban, Self-healing polymeric materials, *Chem. Soc. Rev.*, 2013, **42**(17), 7446–7467.
23. M. Caruso, D. A. Davis, Q. Shen, S. A. Odom, N. R. Sottos, S. R. White and J. S. Moore, Mechanically-Induced Chemical Changes in Polymeric Materials, *Chem. Rev.*, 2009, **109**(11), 5755–5758.
24. F. R. Jones, W. Zhang, M. Branthwaite and F. R. Jones, Self-healing of damage in fiber-reinforced polymer-matrix composites, *J. R. Soc., Interface*, 2007, **4**(13), 381–387.
25. S. D. Bergman and F. Wudl, Mendable Polymers, *J. Mater. Chem.*, 2008, **18**, 41–62.
26. S. R. White, N. R. Sottos, P. H. Geubelle, J. S. Moore, M. R. Kessler, S. R. Sriram, E. N. Brown and S. Viswanathan, Autonomic healing of polymer composites, *Nature*, 2001, **409**(6822), 794–797.
27. J. W. C. Pang and I. P. Bond, A Hollow Fibre Reinforced Polymer Composite Encompassing Self-Healing and Enhanced Damage Visibility, *Compos. Sci. Technol.*, 2005, **65**(11–12), 1791–1799.
28. S. J. Kalista Jr., Self-Healing of Thermoplastic Poly(Ethylene-co-Methacrylic Acid) Copolymers Following Projectile Puncture, MSc Thesis, Virginia Polytechnic Institute, and State University, Mechanical Engineering, September 2003.
29. K. Pingkarawat, C. Dell'Olio, R. J. Varley and A. P. Mouritz, Poly(ethylene-co-methacrylic acid) (EMAA) as an efficient healing agent for high-performance epoxy networks using diglycidyl ether of bisphenol A (DGEBA), *Polymer*, 2016, **92**, 153–163.
30. R. S. McLean, M. Doyle and B. B. Sauer, High-Resolution Imaging of Ionic Domains and Crystal Morphology in Ionomers Using AFM Techniques, *Macromolecules*, 2000, **33**(17), 6541–6550.
31. A. Fallahi, Y. Bahramzadeh, E. Tabatabaie and M. Shahinpoor, A Novel Multifunctional Soft Robotic Transducer Made with Poly (Ethylene-co-Methacrylic Acid) Ionomer Metal Nanocomposite, *Int. J. Intell. Robot. Appl.*, 2017, DOI: 10.1007/s41315-017-0013-y.
32. A. Stewart, The 'living concrete' that can heal itself, http://www.cnn.com/2015/05/14/tech/bioconcrete-delft-jonkers/index.html.

24 Overview of Janus Particles as Smart Materials

Shan Jiang* and Kyle Miller

Materials Science and Engineering, Iowa State University, Ames 50014, USA
*Email: sjiang1@iastate.edu

24.1 Introduction

In ancient Roman times, *Janus* was the god who had two faces (beginnings and endings) (Figure 24.1a). In modern science, we have adapted the term to describe particles with two distinct and usually contrasting sides. These particles have the resemblance of the Taijitu symbol in ancient Asian philosophy (Figure 24.1b), where Yin and Yang (dark and bright) were used to describing seemingly opposite forces. It is believed that these two basic elements give rise to complicated change and transition in the whole world. In the same sense, Janus particles are defined by their duality (Figure 24.1c and d), which can take on a variety of forms and create a wide range of new materials with the simple Janus motif.

The possibilities for properties that can be assigned to each half of the Janus particles are vast (for example, hydrophobicity and charge), and are limited only by the fabrication capabilities of their creators.

Furthermore, the particle geometry is not limited to a sphere with two equal hemispheres; Janus rods, dumbbells, and sheets are structures that exhibit unique properties in their own right, and the ratio of the two surfaces does not necessarily need to be 1:1. As a

Fundamentals of Smart Materials
Edited by Mohsen Shahinpoor
© The Royal Society of Chemistry 2020
Published by the Royal Society of Chemistry, www.rsc.org

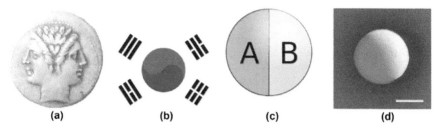

Figure 24.1 (a) Image of the god Janus engraved on an ancient Roman coin, (b) Taoist Taijitu symbol used on the flag of South Korea, (c) diagram showing a Janus particle with two distinct faces, and (d) is an SEM image of a Janus particle, where the scale bar is 500 nm. The bright side is gold, the gray is silica.

result of these features, Janus particles can self-assemble into many unique structures. Also, when one surface responds to an external field, Janus particles become smart materials that are capable of adapting their assembled structures to environmental changes.

This chapter will provide an overview of the properties and applications of Janus particles as smart materials. To do this effectively, we must first go over some of the common strategies for fabricating Janus materials, since fabrication is still the primary limitation by which combinations of properties can be achieved. Then, we will delve into the properties that emerge from these combinations, citing several examples of the self-assembled structures demonstrated by Janus particles. Finally, we will look at some potential applications for Janus systems.

24.2 History and Fabrication of Janus Particles

The first publication on Janus particles was by Casagrande and Veyssie in 1988,[1] and was notably mentioned by Pierre-Gilles de Gennes in his Nobel Prize speech in 1991.[2] While discussing the emerging field of soft matter, he also described taking advantage of Janus particle adsorption at the air–water interface to create a 'skin that can breathe'. The key blockade in the development of Janus particles was their fabrication – it was simply too difficult to get distinct chemistries onto a well-defined geometry on a small colloidal particle. In the early 2000s, Janus particle fabrication methods started to emerge. Since then, the number of articles published on Janus particles has increased dramatically year on year, a trend that continues even today.[3]

One of the earliest methods for producing Janus particles was *via* the directional coating of a monolayer of particles. Although the quantity produced is limited, this technique offers a straightforward

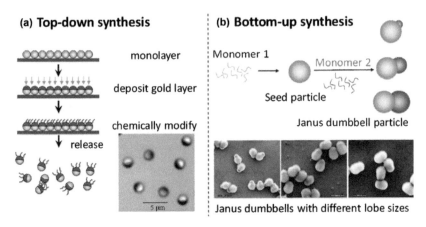

Figure 24.2 (a) Top-down and (b) bottom-up approaches of fabricating Janus particles.

way to fabricate Janus particles with well-defined geometry. To begin with, core particles are synthesized *via* conventional means and made into a monolayer (Figure 24.2a). Then, a coating material is deposited onto the top surface, producing a particle with two different and equally sized hemispheres. The same procedure can be carried out with a wide variety of coating materials; common examples include gold, which is easily modified by thiol chemistry, or nickel, which is magnetic in even nanometer-thin layers. While this technique is simple and versatile, it does not scale to bulk quantities easily, due to the limited yield of the monolayer. Its output is typically limited to milligrams per batch.

The directional coating can be described as a 'top-down' technique, where the particles are already formed and then modified *via* a physical deposition process. Alternative fabrication methods address the scalability issue by using various 'bottom-up' techniques, where Janus particles are self-assembled or formed together from individual atoms or ions. One example is nucleating a new particle onto the surface of the old one, and further phase separation creates a dumb-bell geometry in which each lobe can be adjusted almost independ-ently. The 'bottom-up' method requires careful control of the reaction conditions since the phase separation process is sensitive to the re-action environment. However, this method can potentially produce large quantities of Janus particles with well-defined geometry.

One key concept used to describe the geometry of Janus particles is *Janus balance*, which qualitatively corresponds to the relative areas of the two sides on a Janus particle. A more rigorous, quantified defin-ition has also been developed based on the adsorption energy of Janus

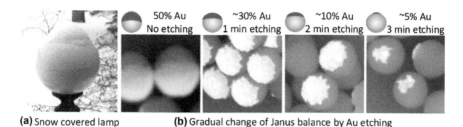

(a) Snow covered lamp **(b)** Gradual change of Janus balance by Au etching

Figure 24.3 Janus balance gradually changed by etching of Au on coated Janus particles.

particles at an interface.[4] A simple method to gradually change Janus balance in Au-coated Janus particles is by etching away the Au coating (Figure 24.3).

The above examples are by no means an exhaustive list but do highlight some of the innovative techniques used to produce Janus particles. It is important to consider a variety of factors before deciding which methods are selected to fabricate Janus particles. For example, a deposition technique can produce homogeneous Janus particles (assuming a perfect monolayer) while a phase separation method is more stochastic in nature and harder to control. When considering Janus particles for a particular application, some questions to ask are: how fine a control is needed? What surface chemistry do you need to create? How easy is it to adjust the Janus balance? And of course, how many particles are needed and how expensive will they be? The first step towards using Janus particles as smart materials is creating them. While we do not intend to go into further detail in this chapter, we encourage an interested reader to look into some of the reviews and book chapters referenced here.[3,5–7]

24.3 Self-assembly Structures

The most distinctive property of Janus particles is how their anisotropic interactions with each other can lead to unique assembled structures. One way to think about the assembled structure is to compare Janus particles and small surfactant molecules. Surfactants adsorb strongly at the interface and assemble into small micelles. Amphiphilic Janus particles behave very similarly as they also aggregate at interfaces,[8] and assemble into well-defined clusters.[9] Also, Janus particles can be manipulated by external fields.[10] It is important to note that it is not just the dual functionalities that enable these behaviors, but the anisotropy of the distribution of the functionalities on the particle surface.

A classic example is the assembly behavior of amphiphilic (one side hydrophobic, the other hydrophilic) Janus particles suspended in an aqueous solution. Adding salt weakens the electrostatic repulsion between anisotropic Janus particles, and they assemble into unique and well-defined structures (Figure 24.4),[9,11] allowing observation of different sized particle clusters. Initially, small particle clusters are formed (Figure 24.4a and 24.4b). When the number of particles becomes large (>7), long chains are formed (Figure 24.4c). At an even higher concentration, under 2D confinement by gravity, amphiphilic Janus particles form unique crystal structures (Figure 24.4d). Computational modeling revealed the detailed structures and the schematic plots (Figure 24.4e and 24.4f) clearly indicate that the hydrophobic sides of the Janus particles prefer to orient towards each other during assembly (similar to small surfactant micelle structures).[9]

While self-assembly without external input is important and useful, when Janus particles are made as smart materials, they can respond to environmental changes and create even more interesting structures.

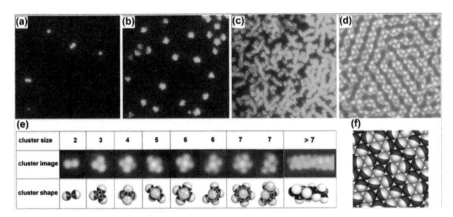

Figure 24.4 Assembled structures (a)–(c) as 3D structures as the particle concentration increases; (d) 2D structures; (e) clusters of different sizes; (f) Janus particle packing in a 2D crystal structure.
Parts a–c, and e are reproduced from ref. 9 with permission from AAAS, Copyright 2011. Parts d and f are reproduced from ref. 11, https://doi.org/10.1103/PhysRevLett.112.218301, with permission from the American Physical Society, Copyright 2014.

24.4 Structure and Motion of Janus Particles Under an External Field

24.4.1 Assembly of Magnetic Janus Particles Under a Magnetic Field

Let us first consider Janus particles coated with magnetic materials on one side. As one would expect, these particles, typically made up of silica cores and a nickel coating,[12] display an anisotropic magnetic response, making them controllable by an applied field. In early experiments, these magnetized particles were examined in a pseudo-2D environment and were shown to aggregate into chains.[13] The particles first aligned with their magnetic hemispheres facing the same direction. Then the spheres formed lines of particles, and similar strings of particles quickly attached, forming chains two particles wide (Figure 24.5). The rotation of the chains was also controlled by the orientation of a magnetic field.

When the particle geometry is changed from spherical to rod-shaped, flexible Janus ribbons can be formed instead,[14] as shown in Figure 24.6. Note the similarities between the side view shown in Figure 24.6c (left panel) and the simple chains in Figure 24.5 – both align particles into lines, but the long axis of the rods allows a different type of stacking.

When the field is shifted in both the spherical and rod-shaped systems, the particles shift as well. However, the rod-shaped Janus particles use this motion to lower their total energy by folding into one of several types of rings. These rings are stable even without the

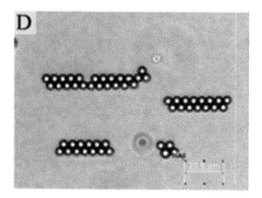

Figure 24.5   Magnetic Janus particles aggregating into chains. Adapted from ref. 13 with permission from the Royal Society of Chemistry.

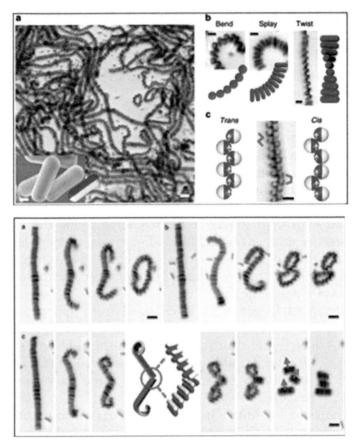

Figure 24.6   Top: magnetised rods will naturally form ribbons. Bottom: when the field is reversed, the ribbons collapse.
Adapted from ref. 14 with permission from Springer Nature, Copyright 2013.

applied field, but they are able to transition back into ribbons under a very strong magnetic field. This transition is not instantaneous and can be stopped to produce a variety of transitional states with partially unfolded rings. A final example of the possibilities of magnetic Janus particles uses an oscillating magnetic field rather than a static one to produce microtubules of Janus particles.[10] As particles rotate in sync with the field, they begin to synchronize their motion with each other, so that neighbor particles take on a complimentary orientation to each other. This process of synchronization is similar to how applause of a large audience becomes rhythmic after a few moments and develops pulses. Under certain configurations, this complementary orientation results in the development of a Janus particle

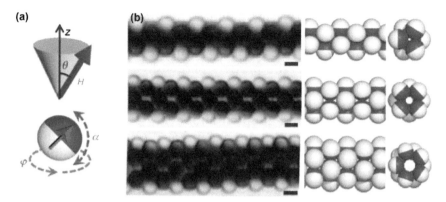

Figure 24.7 (a) Diagram showing dynamic field oscillation axes and (b) microtubule images.
Adapted from ref. 10 with permission from Springer Nature, Copyright 2012.

microtubule, as shown in Figure 24.7. Not shown in the picture is that these microtubules are constantly spinning, yet remain stable.

Magnetic systems are attractive for assembly purposes due to the easy manipulation of the direction and strength of the forces using external fields. Complex structures can assemble from relatively basic particles, and the structures and techniques shown so far can be generalized to other types of magnetic systems.

In addition to what is shown here, rods have been constructed with end-to-end aggregation,[15] and particles have been manipulated into several other structures using magnetic Janus particles.[16–18] Multi-functional particles such as core–shell can also contain combinations of properties, but it is key to note again that, in Janus systems, the definitive enabling property is the anisotropy of the interactions between the particles.

24.4.2 Motion and Assembly of Metal-coated Janus Particles Under an Electric Field

One application of Janus particles is to use their directionality to make particles move. These self-propelling particles rely on some field, gradient, or chemical reaction that acts differently on the two sides of a Janus particle, and produces a net acceleration in one direction. This property stands in contrast to magnetic particle assembly, in which particles are largely manipulated to form structures based on interactions between magnetic components themselves.

A model system for self-propelling particles relies on electro-phoresis to move a metal-coated Janus particle. An AC electric field induces a charge in the metallic half, which creates flows in the solution that push the particle perpendicular to the field axis, shown clearly in Figure 24.8b. This phenomenon was first explored by Gangwal *et al.* in 2008, where they expanded on their experi-mental observations and adopted a model (Figure 24.8c) to explain this effect (Figure 24.8a), termed as induced-charge electrophoresis (ICEP).[19] Notably, the experimental results showed a quadratic relationship between field and particle speed, and that salt affecting the electric double layer also had a strong effect on electrophoresis.

Later, Yan *et al.* demonstrated that ICEP could be used to assemble Janus particles into configurations of active clusters.[20] They used an electric field to manipulate the relative charge strength on the two sides of the particles by using an AC electric

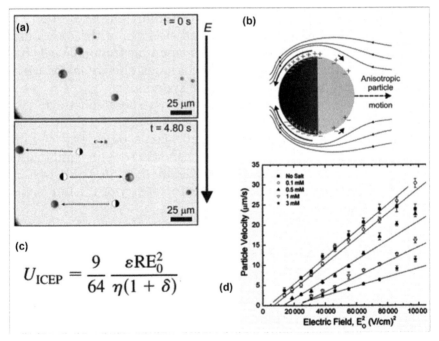

Figure 24.8   (a) Microscope images showing ICEP in action; (b) an illus-trated model of how ICEP works; (c) the defining equation for ICEP energy; (d) a graph demonstrating the effect of electric field strength and salt concentration on particle velocity. Adapted from ref. 19, https://doi.org/10.1103/PhysRevLett.100. 058302, with permission from the American Physical Society, Copyright 2008.

field of controllable frequency. At low frequencies, no charge is induced, so the particles move randomly. As the frequency increases, a weak charge is induced on the metallic half of the Janus particles, and they start to repel each other until their motion aligns. The particles in this regime behave like a swarm and have interesting wave-like behavior. At higher frequencies, the emerging charge matches the original charge on the other side, which causes the particles to lock into chains and flow past each other in lines. At the highest frequency tested, the newly formed charge is strong enough that it propels the particles into each other, and large, dynamic clusters form. Theoretical models and computer simulations were created to explain this behavior, and it was tested thoroughly in a pseudo-2D environment. At every stage, the computer models agree with the experimental results, reaffirming the role charge plays in electrophoresis.

It is important to note that with a homogenous particle, one can still manipulate the surface to have varying levels of charge. However, it is the physical separation of the charges on a single Janus particle that results in this unique behavior. The Janus geometry enables assembly in every example shown so far.

Particle motion is, of course, not limited to electrophoresis. A variety of methods have emerged over recent years, using light,[21] chemical reactions,[22] and induced gradients[23] to develop self-propelling particles. This area of Janus research has the potential to unlock promising applications. In one recent example, these particles are being used to pierce through biological barriers that would otherwise prevent drug delivery.[23]

So far, this chapter has limited itself to discussing the fundamentals of Janus particle assembly and providing highlights of how particles can assemble under varying conditions. However, as Janus particle research evolves, researchers are looking into potential applications and taking advantage of the unique properties of Janus particles. Janus particles do have several notable functions, such as an ability to stabilize emulsions,[5] serve as biphasic catalysts,[24] or act as reconfigurable materials.[20] Figure 24.9 shows that Janus particles adsorb strongly at interfaces and can stabilize emulsions for an extended period.

Another promising area of Janus particle research is in biological applications. The field of biomedicine has been intensely researched, and every aspect of a particle can influence its effectiveness as a drug delivery vehicle. There is a need for precise control over particle parameters to determine *in vivo* behavior. This becomes

Figure 24.9   (a) Janus particles adsorbed at an interface; (b) photos of water-in-oil emulsions stabilized by Janus particles for an extended period.
Adapted from ref. 5 with permission from John Wiley and Sons, Copyright © 2010 WILEY-VCH Verlag GmbH & Co. KGaA, Weinheim.

more challenging as the properties of materials are correlated – often, the enhancement of one property negatively affects others. Janus particles offer a simple and elegant solution: compartmentalize different properties into one particle. The resulting system will have not only the desired multi-functionality, but also an anisotropy that can be used to enhance its applications even further.

A simple demonstration of Janus particles in this system is a recent study on T cell activation with magnetic Janus particles.[25] The idea was to use a simple bar magnet to manipulate a single functional particle so that it would activate a T cell. The experiment used 3 µm silica particles with a layer of nickel and aluminum for magnetic and oxidation protection, respectively. The silica hemisphere was then functionalized by adding anti-CD3 antibodies, which cause T cell activation. It was demonstrated that magnetic Janus particles could be used to activate a T cell remotely by simply rotating the particles under an external magnetic field. The T cell gave off a clear calcium ion signal when activated (Figure 24.10).

As another example, Janus particles can offer potential solutions in the coating industry. Since amphiphilic molecules are widely used in coating formulations, Janus particles may be applied to replace small surfactant or dispersant molecules and improve the coating performance.[26] However; it remains challenging to synthesize Janus particles economically for large-scale industrial applications.

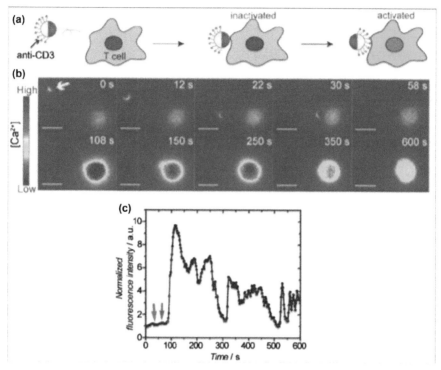

Figure 24.10 (a) Schematic plot showing an overview of the procedure; (b) Fluorescence images showing a particle (the crescent in the top left) and reaction in a T-cell (larger light spot); (c) graph of the total fluorescence, with the blue arrow indicating where the particle made contact.
Adapted from ref. 26 with permission from John Wiley and Sons, © 2016 WILEY-VCH Verlag GmbH & Co. KGaA, Weinheim.

24.5 Conclusions

Janus particles represent a leap forward for colloidal science. Their properties are more than just the sum of their parts; their anisotropy represents a new dimension in colloid science that enables a variety of new assembled structures. The interactions between these particles can produce unique and controllable structures that are simply impossible to produce with homogenous particles. As more and more synthetic methods for Janus particles are being developed, Janus particles have begun to find more applications in different areas of research. However, additional work is still needed to link the fundamental structures formed by Janus particles with the material performance of the final product.

Homework Problems

Homework Problem 24.1

Compare the Janus balance with the hydrophile–lipophile balance (HLB) of the surfactant molecules, what are the similarities?

Homework Problem 24.2

Find a recent paper that uses Janus particles as smart materials. Briefly describe why Janus particles are different from homogeneous particles in that system? What value does being Janus add to the system?

Homework Problem 24.3

Compare and contrast top-down and bottom-up methods for producing Janus particles in terms of particle homogeneity, versatility (what limits which materials can be used?), and scalability.

Homework Problem 24.4

Read ref. 27, where *Janus balance* is defined as the dimensionless ratio of work to transfer a Janus particle from the oil–water interface into the oil phase, normalized by the work needed to move it into the water phase. In a common case, the Janus balance can be calculated using the following equation:

$$J = \frac{\sin^2 \alpha + 2 \cos \theta_p (\cos \alpha - 1)}{\sin^2 \alpha + 2 \cos \theta_a (\cos \alpha + 1)}$$

where α is determined by the Janus geometry and θ_a and θ_p correspond to the contact angles of the hydrophobic and hydrophilic sides, as shown in the Figure 24.11.

(a) What is the Janus balance value when particles have the maximum adsorption energy?

(b) Calculate the Janus balance, J, given that θ_a, θ_p and α in this system are $0°$, $121°$, and $90°$.

(c) Calculate the value of α that will give the maximized adsorption energy, given the same values of θ_a and θ_p in (b).

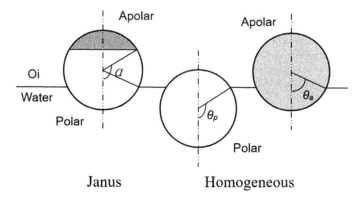

Figure 24.11 Janus geometry.

References

1. C. Casagrande and M. Veyssie, Janus beads – realization and 1st observation of interfacial properties, *C. R. Acad. Sci., Ser. II*, 1988, **306**(20), 1423–1425.
2. P. G. Degennes, Soft matter (nobel lecture), *Angew. Chem., Int. Ed. Engl.*, 1992, **31**(7), 842–845.
3. Z. Yang, A. H. Muller, C. Xu, P. S. Doyle, J. M. DeSimone, J. Lahann, F. Sciortino, S. Glotzer, L. Hong and D. A. Aarts, *Janus Particle Synthesis, Self-assembly, and Applications*, Royal Society of Chemistry, 2012.
4. S. Jiang and S. Granick, Janus balance of amphiphilic colloidal particles, *J. Chem. Phys.*, 2007, **127**, 16.
5. S. Jiang, Q. Chen, M. Tripathy, E. Luijten, K. S. Schweizer and S. Granick, Janus Particle Synthesis and Assembly, *Adv. Mater.*, 2010, **22**(10), 1060–1071.
6. S. G. Shan Jiang, *Janus Particle Synthesis, Self-Assembly, and Applications*, Royal Society of Chemistry, 2013, p. 312.
7. J. Zhang, B. A. Grzybowski and S. Granick, Janus Particle Synthesis, Assembly, and Application, *Langmuir*, 2017, **33**(28), 6964–6977.
8. N. Glaser, D. J. Adams, A. Böker and G. Krausch, Janus particles at liquid–liquid interfaces, *Langmuir*, 2006, **22**(12), 5227–5229.
9. Q. Chen, J. K. Whitmer, S. Jiang, S. C. Bae, E. Luijten and S. Granick, Supracolloidal reaction kinetics of Janus spheres, *Science*, 2011, **331**(6014), 199–202.
10. J. Yan, M. Bloom, S. C. Bae, E. Luijten and S. Granick, Linking synchronization to self-assembly using magnetic Janus colloids, *Nature*, 2012, **491**(7425), 578.
11. S. Jiang, J. Yan, J. K. Whitmer, S. M. Anthony, E. Luijten and S. Granick, Orientationally Glassy Crystals of Janus Spheres, *Phys. Rev. Lett.*, 2014, **112**(21), 218301.
12. I. Sinn, P. Kinnunen, S. N. Pei, R. Clarke, B. H. McNaughton and R. Kopelman, Magnetically uniform and tunable Janus particles, *Appl. Phys. Lett.*, 2011, **98**(2), 024101.
13. S. K. Smoukov, S. Gangwal, M. Marquez and O. D. Velev, Reconfigurable responsive structures assembled from magnetic Janus particles, *Soft Matter*, 2009, **5**(6), 1285–1292.
14. J. Yan, K. Chaudhary, S. Chul Bae, J. A. Lewis and S. Granick, Colloidal ribbons and rings from Janus magnetic rods, *Nat. Commun.*, 2013, **4**, 1516.
15. K. Chaudhary, J. J. Juárez, Q. Chen, S. Granick and J. A. Lewis, Reconfigurable assemblies of Janus rods in AC electric fields, *Soft Matter*, 2014, **10**(9), 1320–1324.

16. S. Sacanna, L. Rossi and D. J. Pine, Magnetic click colloidal assembly, *J. Am. Chem. Soc.*, 2012, **134**(14), 6112–6115.
17. J. Yan, S. C. Bae and S. Granick, Colloidal superstructures programmed into magnetic Janus particles, *Adv. Mater.*, 2015, **27**(5), 874–879.
18. A. Snezhko and I. S. Aranson, Magnetic manipulation of self-assembled colloidal asters, *Nat. Mater.*, 2011, **10**(9), 698.
19. S. Gangwal, O. J. Cayre, M. Z. Bazant and O. D. Velev, Induced-charge electrophoresis of metallodielectric particles, *Phys. Rev. Lett.*, 2008, **100**(5), 058302.
20. J. Yan, M. Han, J. Zhang, C. Xu, E. Luijten and S. Granick, Reconfiguring active particles by electrostatic imbalance, *Nat. Mater.*, 2016, **15**(10), 1095.
21. M. Xuan, Z. Wu, J. Shao, L. Dai, T. Si and Q. He, Near-infrared light-powered Janus mesoporous silica nanoparticle motors, *J. Am. Chem. Soc.*, 2016, **138**(20), 6492–6497.
22. D. P. Singh, U. Choudhury, P. Fischer and A. G. Mark, Non-Equilibrium Assembly of Light-Activated Colloidal Mixtures, *Adv. Mater.*, 2017, **29**(32), 1701328.
23. A. Joseph, C. Contini, D. Cecchin, S. Nyberg, L. Ruiz-Perez, J. Gaitzsch, G. Fullstone, X. Tian, J. Azizi and J. Preston, Chemotactic synthetic vesicles: Design and applications in blood-brain barrier crossing, *Sci. Adv.*, 2017, **3**(8), e1700362.
24. L. Baraban, D. Makarov, R. Streubel, I. Monch, D. Grimm, S. Sanchez and O. G. Schmidt, Catalytic Janus motors on microfluidic chip: deterministic motion for targeted cargo delivery, *ACS Nano*, 2012, **6**(4), 3383–3389.
25. K. Lee, Y. Yi and Y. Yu, Remote control of T cell activation using magnetic Janus particles, *Angew. Chem.*, 2016, **128**(26), 7510–7513.
26. S. Jiang, A. Van Dyk, A. Maurice, J. Bohling, D. Fasano and S. Brownell, Design colloidal particle morphology and self-assembly for coating applications, *Chem. Soc. Rev.*, 2017, **46**(12), 3792–3807.
27. S. Jiang and S. Granick, *Chem. Phys.*, 2007, **127**, 161102.

Appendix

This appendix lists the Royal Society of Chemistry's books on smart materials, including a synopsis of each volume. Books in the Smart Materials Series cover fundamentals and applications of different material systems from renowned international experts. The series is edited by Hans-Jörg Schneider, Saarland University, Germany and Mohsen Shahinpoor, University of Maine, USA. Books in the series are listed in the ISI Books Citation Index and Scopus, and all titles can be found on the Royal Society of Chemistry website at rsc.li/smart-materials.

Intelligent Materials

Editors: Mohsen Shahinpoor and Hans-Jörg Schneider
ISBN: 978-0-85404-335-4, 2007

Synopsis

In this exceptional text, the expertise of specialists across the globe is drawn upon to present a truly interdisciplinary outline of the topic. The influence of current research in this field on future technology is undisputed, and potential applications of intelligent materials span nanoscience, nanotechnology, medicine, engineering, biotechnology, pharmaceutical, and many other industries. This is an authoritative introduction to the most recent developments in the area, which will provide the reader with a better understanding of the almost un-limited opportunities in the progress and design of new intelligent

Fundamentals of Smart Materials
Edited by Mohsen Shahinpoor
© The Royal Society of Chemistry 2020
Published by the Royal Society of Chemistry, www.rsc.org

materials. An indispensable reference for anyone contemplating working in the field including a foreword from distinguished scientist and Nobel laureate P.-G de Gennes: "This will be the starting point for all researchers looking for industrial solutions involving smart materials. Congratulations to the Editors for providing such a vast and interdisciplinary book." P.-G de Gennes, France Prix Nobel de Physic 1991.

Janus Particle Synthesis, Self-Assembly and Applications

Editors: Shan Jiang and Steve Granick
ISBN: 978-1-84973-423-3, 2012

Synopsis

Named after the two-faced Roman god, Janus particles have gained much attention due to their potential in a variety of applications, including drug delivery. This is the first book devoted to Janus particles and covers their methods of synthesis, how these particles self-assemble, and their possible uses. By following the line of synthesis, self-assembly, and applications, the book not only covers the fundamental and applied aspects, but it goes beyond a simple summary and offers a logistic way of selecting the proper synthetic route for Janus particles for certain applications. Written by pioneering experts in the field, the book introduces the Janus concept to those new to the topic and highlights the most recent research progress on the topic for those active in the field and catalyze new ideas.

Smart Materials for Drug Delivery: Complete Set of Two Volumes

Editor(s): Carmen Alvarez-Lorenzo and Angel Concheiro
ISBN: 978-1-84973-552-0, 2013

Synopsis

Smart materials, which can change properties when an external stimulus is applied, can be used for the targeted drug delivery of an active molecule to a specific site in the correct dosage. Different materials such as liposomes, polymeric systems, nanomaterials, and hydrogels can respond to different stimuli such as pH, temperature, and light, and these are all attractive for controlled release applications. With so many papers available on smart and stimuli-responsive materials for drug delivery applications, it's hard to know where to start reading about this exciting topic. In two volumes, **Smart Materials for Drug Delivery** brings together the recent findings in the area and provides a critical analysis of the different materials available and how they can be applied to advanced drug delivery systems. With contributions from leading experts in the field, including a foreword from distinguished scientist Nicholas Peppas, The University of Texas at Austin, USA, the book will provide both an introduction to the key areas for graduate students and new researchers in the stimuli-responsive field as well as serving as a reference for those already working on fundamental materials research or drug delivery applications.

Materials Design Inspired by Nature: Function Through Inner Architecture

Editors: Peter Fratzl, John W. C. Dunlop and Richard Weinkamer
ISBN: 978-1-84973-553-7, 2013

Synopsis

The inner architecture of a material can have an astonishing effect on its overall properties and is vital to understand when designing new materials. Nature is a master at designing hierarchical structures, and so researchers are looking at biological examples for inspiration, specifically to understand how nature arranges the inner architectures for a particular function to apply these design principles into human-made materials. **Materials Design Inspired by Nature** is the first book to address the relationship between the inner architecture of natural materials and their physical properties for materials design. The book explores examples from plants, the marine world, arthropods and bacteria, where the inner architecture is exploited to obtain specific mechanical, optical or magnetic properties along with how these design principles are used in human-made products. Details of the experimental methods used to investigate hierarchical structures are also given. Written by leading experts in bio-inspired materials research, this is essential reading for anyone developing new materials.

Responsive Photonic Nanostructures: Smart Nanoscale Optical Materials

Editor: Yadong Yin
ISBN: 978-1-84973-653-4, 2013

Synopsis

Photonic crystal nanostructures, whose photonic properties can be tuned in response to external stimuli, are desired for a wide range of applications in color displays, biological and chemical sensors, and inks and paints. Until now, there is no single resource which gives a complete overview of these exciting smart materials. **Responsive Photonic Nanostructures: Smart Nanoscale Optical Materials** details the fabrication of photonic crystal structures through self-assembly approaches, general strategies, and approaches for creating responsive photonic structures for different responsive systems such as chemical, optical, electrical and magnetic as well as their applications. With contributions from leading experts in the field, this comprehensive summary on Responsive Photonic Nanostructures is suitable for postgraduates and researchers in academia and industry interested in smart materials and their potential applications.

Magnetorheology: Advances and Applications

Editor: Norman M. Wereley
ISBN: 978-1-84973-667-1, 2013

Synopsis

Magnetorheological fluids, smart fluids which change viscosity in the presence of a magnetic field, are of great commercial interest for many engineering applications such as shock absorbers and dampers in aerospace. **Magnetorheology: Advances and Applications** provides an update on the key developments in the physics, chemistry, and uses of magnetorheological fluids. Topics covered include the role of interparticle friction and rotational diffusion, magnetoelasticity, nondimensional flow analysis, thin-film rheology, tribology, coated magnetorheological composite particles and magnetorheological devices with multiple functions. Specific chapters on applications cover adaptive magnetorheological energy absorbing mounts for shock mitigation, magnetorheological fluid-based   high   precision   finishing   technologies,   adaptive magnetorheological landing gear systems, and magnetorheological lag dampers for stability augmentation in helicopters. Edited by a leading expert and with contributions from distinguished scientists in the field, this timely book is suitable for chemists, physicists, and engineers wanting to gain a comprehensive overview of these smart materials.

Mechanochromic Fluorescent Materials: Phenomena, Materials and Applications

Editors: Jiarui Xu and Zhenguo Chi
ISBN: 978-1-84973-821-7, 2014

Synopsis

Mechanochromic fluorescent (or mechanofluorochromic) materials change their emission colors (spectra) when an appropriate external mechanical force stimulus is applied. This is an important group of materials with a huge range of applications, including use in sensors, memory chips, security inks, and light devices. **Mechanochromic Fluorescent Materials** introduces the reader to the concept of mechanofluorochromism and the variety of applications of this group of materials. Prominent international figures in mechano-fluorochromism consider innovative research in this field over the last ten years. Chapters provide in-depth coverage of most reported mechanofluorochromic systems, including organic and organic-inorganic complexes; polymer and polymer composites; and aggregation-induced emission. This book is aimed to inform all students and researchers with interest in mechanofluorochromism, and to help researchers identify and synthesize more of these materials, and develop the study and application of mechanofluorochromic materials.

Cell Surface Engineering: Fabrication of Functional Nanoshells

Editors: Rawil Fakhrullin, Insung Choi and Yuri Lvov
ISBN: 978-1-84973-902-3, 2014

Synopsis

Cell surface engineering is an emerging field concerning cell surface modifications to enhance its functionalities. The book introduces the reader to the area of surface-functionalized cells and summarizes recent developments in the area including fabrication, characterization, applications, and nanotoxicity. Topics covered include recent approaches for the functionalization of cells with nanomaterials (polymer nanofilms and nanoparticles), fabrication of functional biomimetic devices and assemblies based on nanoparticle-modified microbial cells and artificial spores (the bioinspired encapsulation of living cells with inorganic nanoshells). The book provides an interdisciplinary approach to the topic with authors from both biological and chemical backgrounds. This multidisciplinary view makes the book suitable for those interested in biomaterials, biochemistry, microbiology and colloid chemistry, providing both an introduction for postgraduate students as well as a comprehensive summary for those already working in the area. Biomaterials, biochemistry, microbiology, and colloid chemistry.

Functional Nanometer-Sized Clusters of Transition Metals: Synthesis, Properties, and Applications

Editors: Wei Chen and Shaowei Chen
ISBN: 978-1-84973-824-8, 2014

Synopsis

Metal nanoclusters, which bridge metal atoms and nanocrystals, are gaining attention due to their unique chemical and physical properties which differ greatly from their corresponding large nanoparticles and molecular compounds. Their electronic and optical properties are of particular interest for their use in sensing, optoelectronics, photovoltaics, and catalysis. The book highlights recent progress and challenges in size-controlled synthesis, size-dependent properties, characterization and applications of metal nanoclusters. Specific topics include organochalcogenolate-stabilized metal nanoparticles, water-soluble fluorescent silver nanoclusters, thiolate-protected Au and Ag nanoclusters, DNA-templated metal nanoclusters, fluorescent platinum nanoclusters and Janus nanoparticles by interfacial engineering. Edited by active researchers in the area, the book provides a valuable reference for researchers in the area of functional nanomaterials. It also provides a guide for graduate students, academic and industrial researchers interested in the fundamentals of the materials or their applications.

Biointerfaces: Where Material Meets Biology

Editors: Dietmar Hutmacher and Wojciech Chrzanowski
ISBN: 978-1-84973-876-7, 2014

Synopsis

To design and develop new biomaterials, it is essential to understand the biointerface, the interconnection between a synthetic or natural material and tissue, microorganism, cell, virus, or biomolecule. **Biointerfaces: Where Material Meets Biology** provides an up to date overview of the knowledge and methods used to control living organism responses to implantable devices. The book starts with an introduction to the biointerface – past, present and the future perspectives and covers the key areas of biomolecular interface for cell modulation, topographical biointerface, mechano structural bio interface, chemo-structural biointerfaces and interface that control bacteria response. By combining the cellular, antimicrobial, antibacterial and therapeutic aspects of the interface with the methodology of fabrication and testing of the synthetic biomaterials used in a variety of medical applications, the text provides a handbook for researchers. Edited by leading researchers, the book integrates the understanding of cell, microorganism and biomolecule interactions with surfaces and the methods used for assessment that appeals to materials scientists, chemists, biotechnologists, (molecular-) biologists, biomedical engineers interested in the fundamentals and applications of biomaterials and biointerfaces.

Supramolecular Materials for Opto-Electronics

Editor: Norbert Koch
ISBN: 978-1-84973-826-2, 2014

Synopsis

For years, concepts and models relevant to the fields of molecular electronics and organic electronics have been invented in parallel, slowing down progress in the field. This book illustrates how synthetic chemists, materials scientists, physicists, and device engineers can work together to reach their desired, shared goals, and provides the knowledge and intellectual basis for this venture. **Supramolecular Materials for Opto-Electronics** covers the basic principles of building supramolecular organic systems that fulfill the requirements of the targeted optoelectronic function; specific material properties based on the fundamental synthesis and assembly processes, and provides an overview of the current uses of supramolecular materials in optoelectronic devices. To conclude, a "what's next" section provides an outlook on the future of the field, outlining the ways overarching work between research disciplines can be utilized. Postgraduate researchers and academics will appreciate the fundamental insight into concepts and practices of supramolecular systems for optoelectronic device integration.

Photocured Materials

Editors: Atul Tiwari and Alexander Polykarpov
ISBN: 978-1-78262-001-3, 2014

Synopsis

Traditionally, most synthetically developed materials are hardened by heating them to an elevated temperature, a process requiring large amounts of energy and space. Interest in photocured materials using UV-light is growing due to simplifications in manufacturing and growing environmental concerns; it is expected photocuring could reduce electricity consumption by 90% compared to traditional curing. Photocured materials also reduce evaporation of volatile organic components, curing time and waste, thereby enhancing productivity and reducing workspace. The materials technologies based on photocuring are gaining momentum, and this will be the first book to provide an in-depth focus on the subject. This book summarizes the fundamentals required to understand the field, characterizes the use of novel materials and the development of synthetic aspects, and discusses the future of the technology. The comprehensive review chapters are suitable for a broad readership from diverse backgrounds including chemistry, physics, materials science and engineering, medical science, pharmacy, biotechnology, and biomedical engineering. **Photocured Materials** will be of interest to students, researchers, scientists, engineers, and professors.

Semiconductor Nanowires: From Next-Generation Electronics to Sustainable Energy

Editors: Wei Lu and Jie Xiang
ISBN: 978-1-84973-815-6, 2014

Synopsis

Semiconductor nanowires were initially discovered in the late 1990s and since then there has been an explosion in the research of their synthesis and understanding of their structures, growth mechanisms, and properties. The realization of their unique electrical, optical and mechanical properties has led to a great interest in their use in electronics, energy generation, and storage. This book provides a timely reference on semiconductor nanowires including an introduction to their synthesis and properties and specific chapters focusing on the different applications including photovoltaics, nanogenerators, transistors, biosensors, and photonics. This is the first book dedicated to **Semiconductor Nanowires** and provides an invaluable resource for researchers already working in the area as well as those new to the field. Edited by leading experts in the field and with contributions from well-known scientists, the book will appeal to both those working on fundamental nanomaterial research and those commercially interested in their applications.

Chemoresponsive Materials: Stimulation by Chemical and Biological Signals

Editor: Hans-Jorg Schneider
ISBN: 978-1-78262-062-4, 2015

Synopsis

Smart materials stimulated by chemical or biological signals are of interest for their many applications including drug delivery, as well as in new sensors and actuators for environmental monitoring, process and food control, and medicine. In contrast to other books on responsive materials, this volume concentrates on materials which are stimulated by chemical or biological signals. **Chemoresponsive Materials** introduces the area with chapters covering different responsive material systems including hydrogels, organogels, membranes, thin layers, polymer brushes, chemomechanical and imprinted polymers, nanomaterials, silica particles, as well as carbohydrate- and bio-based systems. Many promising applications are highlighted, with an emphasis on drug delivery, sensors, and actuators. With contributions from internationally known experts, the book will appeal to graduate students and researchers in academia, healthcare, and industry interested in functional materials and their applications.

Functional Metallosupramolecular Materials

Editors: John George Hardy and Felix H. Schacher
ISBN: 978-1-78262-022-8, 2015

Synopsis

There is great interest in metallosupramolecular materials because of their use in magnetic, photonic and electronic materials. **Functional Metallosupramolecular Materials** focuses on the applications of these materials covering the chemistry underlying the synthesis of a variety of ligands to coordinate various metal ions and the generation of 2D and 3D materials based on these constructs. The book starts by looking at different metallosupramolecular systems including naturally occurring functional metallosupramolecular materials; DNA-based metallosupramolecular materials; metallopolymers; metallogels as well as functional materials based on MOFs. Subsequent chapters then systematically cover the different applications such as molecular computation, spin-crossover, light harvesting, and as photocatalysts for the production of solar fuels. The book provides an overview of functional metallosupramolecular materials that will be of interest to graduate students, academics and industrial chemists interested in supramolecular chemistry, materials science, and the materials applications.

Bio-Synthetic Hybrid Materials and Bionanoparticles: A Biological Chemical Approach Towards Material Science

Editors: Alexander Boker and Patrick van Rijn
ISBN: 978-1-84973-822-4, 2015

Synopsis

There is much interest in using biological structures for the fabrication of new functional materials. Recent developments in the particle character and behavior of proteins and viral particles have had a major impact on the development of novel nanoparticle systems with new functions and possibilities. **Bio-Synthetic Hybrid Materials and Bionanoparticles** approaches the subject by covering the basics of disciplines involved as well as recent advances in new materials. The first section of the book focusses on the design and synthesis of different nanoparticles and hybrid structures, including the use of genetic modification as well as by organic synthesis. The second section of the book looks at the self-assembling behavior of nanoparticles to form new materials. The final section looks at bionanoparticle-based functional systems and materials, including chapters on biomedical applications and electronic systems and devices. Edited by leading scientists in nanoparticles, the book is a collaboration between scientists with different backgrounds and perspectives which will initiate the next generation of bio-based structures, materials, and devices.

Ionic Polymer Metal Composites (IPMCs): Smart Multi-Functional Materials and Artificial Muscles, Complete Set

Editor: Mohsen Shahinpoor
ISBN: 978-1-78262-720-3, 2015

Synopsis

Ionic polymer metal composites (IPMCs) can generate a voltage when physically deformed. Conversely, an applied small voltage or electric field, can induce an array of spectacular large deformation or actuation behaviors in IPMCs, such as bending, twisting, rolling, twirling, steering and undulating. An important smart material, IPMCs have applications in energy harvesting and as self-powered strain or deformation sensors, especially suitable for monitoring the shape of dynamic structures. Other uses include soft actuation applications and as a material for biomimetic robotic soft artificial muscles in industrial and medical contexts. This comprehensive set on ionic polymer metal composites provides broad coverage of state of the art and recent advances in the field written by some of the world's leading experts on various characterizations and modeling of IPMCs. The first two chapters cover the fundamentals of IPMCs and methodologies for their manufacture, followed by specific chapters looking at different aspects of actuation and sensing of IPMCs. These include uses in electrochemically active electrodes, electric energy storage devices, soft biomimetic robotics artificial muscles, multiphysics modeling of IPMCs, biomedical applications, IPMCs as dexterous manipulators and tactile sensors for minimally invasive robotic surgery, self-sensing, miniature pumps for drug delivery, IPMC snake-like robots, IPMC microgrippers for microorganisms manipulations, Graphene-based IPMCs and cellulose-based IPMCs or electroactive paper actuators (EAPap). Edited by the leading authority on IMPCs, the broad coverage of this book will appeal to researchers from chemistry, materials, engineering, physics and medical communities interested in both the material and its applications.

Conducting Polymers: Bioinspired Intelligent Materials and Devices

Author: Toribio Fernandez Otero
ISBN: 978-1-78262-315-1, 2015

Synopsis

Conducting polymers are organic, conjugated materials that offer high electrical conductivity through doping by oxidation and a wide range of unique electromechanical and electrochromic characteristics. These properties can be reversibly tuned through electrochemical reactions, making this class of materials good biomimetic models and ideal candidates for the development of novel flexible and transparent sensing devices. This book comprehensively summarises the current and future applications of conducting polymers, with chapters focussing on electrosynthesis strategies, theoretical models for composition dependent allosteric and structural changes, composition dependent biomimetic properties, novel biomimetic devices and future developments of zoomorphic and anthropomorphic tools. Written by an expert researcher working within the field, this title will have broad appeal to materials scientists in industry and academia, from postgraduate level upwards.

Smart Materials for Advanced Environmental Applications

Editor: Peng Wang
ISBN: 978-1-78262-108-9, 2016

Synopsis

The development of smart materials for environmental applications is a highly innovative and promising new approach to meet the increasing demands from society on water resources and pollution remediation. Smart materials with surfaces that can reversibly respond to stimuli from internal and external environments by changing their properties show great promise as solutions for global environmental issues. Many of these functional materials are inspired by biological systems, that use sophisticated material interfaces to display high levels of adaptability to their environment. Leading researchers present the latest information on the current and potential applications of omniphobic slippery coatings; responsive particle stabilized emulsions and self-healing surfaces among other functional materials. The book contains a section dedicated to water treatment and harvesting, describing and explaining strategies such as the use of copolymer membranes and surfaces with patterned wettability. It provides a valuable source of information for environmental, materials, polymer and nano-scientists interested in environmental applications of functional material surfaces.

Self-cleaning Coatings: Structure, Fabrication, and Application

Editor: Junhui He
ISBN: 978-1-78262-286-4, 2016

Synopsis

Recent years have seen fast development in the field of self-cleaning coatings towards varied applications, such as solar cells, flat display panels, smart cellular phones, building windows, oil pipelines, vehicle coatings, and optical devices. The field has been rapidly gaining attention, not only from research and teaching scientists but also from a growing population of college and graduate students. Self-cleaning coatings describe this interesting field, providing details of natural counterparts with self-cleaning functions, theoretical aspects of self-cleaning phenomena, fabrication strategies and methods, applications and industrial impacts. Edited and written by world-renowned scientists in the field, this book will provide an excellent overview of this field and will be of interest to materials and polymer scientists working in industry and academia.

Functional Polymer Composites with Nanoclays

Editors: Yuri Lvov, Baochun Guo and Rawil F. Fakhrullin
ISBN: 978-1-78262-422-6, 2016

Synopsis

Polymer-clay nanocomposites have flame-retardant, antimicrobial, anticorrosion, and self-healing properties; they are biocompatible and environmentally benign. Multiple types of clay minerals may be exfoliated or individually dispersed and then used as natural nanoparticle additives of different size and shape for composite formation. Loading polymers with clays increases their strength. However, it is only recently that such composites were prepared with controlled nanoscale organization allowing for the enhancement of their mechanical properties and functionality. Edited by pioneers in the field, this book will explain the great potential of these materials and will bring together the combined physicochemical, materials science, and biological expertise to introduce the reader to the vibrant field of nanoclay materials. This book will provide an essential text for materials and polymers scientists in industry and academia.

Bioactive Glasses: Fundamentals, Technology, and Applications

Editors: Aldo R. Boccaccini, Delia S. Brauer and Leena Hupa
ISBN: 978-1-78262-976-4, 2016

Synopsis

The global aging society has significantly increased the need for implant materials, which not only replace damaged or lost tissue but are also able to regenerate it. The field of bioactive glasses has been expanding continuously over recent years as they have been shown to bond with hard and soft tissue, release therapeutically active ions, and be capable of enhancing bone formation and regeneration. Also, they are successfully being used to re-mineralize teeth, thereby making bioactive glasses highly attractive materials in both dentistry and medicine. Understanding the multidisciplinary requirements set by the human body's environment and the special characteristics of the different families of bioactive glasses is key in developing new compositions to novel clinical applications. **Bioactive Glasses** aims to bridge the different scientific communities associated with the field of bioactive glasses with a focus on the materials science point of view. Emerging applications covered include soft tissue regeneration, wound healing, vascularisation, cancer treatment, and drug delivery devices. This book provides a comprehensive overview of the latest applications of bioactive glasses for material scientists.

Smart Materials for Tissue Engineering: Two-volume Set

Editor: Qun Wang
ISBN: 978-1-78801-099-3, 2017

Synopsis

In recent years there has been tremendous progress in the area of tissue engineering research. This two-volume set, containing **Smart**

Materials for Tissue Engineering: Fundamental Principles and Smart Materials for Tissue Engineering: Applications, provides a complete overview of the field. Volume one covers the fundamental principles underpinning the materials science developed for enhancing tissue regeneration, as well as those used for regulating the functions of living cells. Volume two focuses on the applications of different materials for replacing or facilitating tissue regeneration. It also provides examples of new materials that have been developed to control cell behaviors and tissue formation by biomimetic topography, which closely replicate the natural extracellular matrix. This set comprehensively documents the recent advancements in smart materials for tissue engineering and provides an essential text for those working in materials science and materials engineering, academia and industry. Please see the webpages for the individual volumes in the set to access the eBook versions.

Magnetic Nanomaterials: Applications in Catalysis and Life Sciences

Editors: Stefan H. Bossmann and Hongwang Wang
ISBN: 978-1-78262-788-3, 2017

Synopsis

Magnetic nanomaterials have undergone a significant evolution during the past decade, with supramolecular nanoparticle organization reaching unprecedented levels of complexity and the materials providing new approaches to treating cancer. **Magnetic Nanomaterials** will provide a comprehensive overview of the latest research in the area of magnetic nanoparticles and their broad applications in synthesis, catalysis, and theranostics. The book starts with an introduction to magnetism in nanomaterials and magnetic nanoparticle design followed by individual chapters which focus on specific uses. Applications covered include drug delivery, theranostic agents for cancer treatment as well as catalysis, biomass conversion, and catalytic enhancement of NMR sensitivity. The reader will have the opportunity to learn about the frontier of magnetic nanotechnology from scientists that have shaped this unique and highly collaborative field of research. Written and edited by experts working within the field across the world, this book will appeal to students and researchers interested in nanotechnology, engineering, and physical sciences.

Biobased Smart Polyurethane Nanocomposites: From Synthesis to Applications

Author: Niranjan Karak
ISBN: 978-1-78801-180-8, 2017

Synopsis

Polyurethane nanocomposites present an attractive and sustainable way for designing smart materials that can be used in packaging, health, and energy applications. **Biobased Smart Polyurethane Nanocomposites** brings together the most recent research in the field from the basic concepts through to their applications. Special emphasis is given to sustainable, biodegradable polyurethane nanocomposites with hyperbranched architecture. The book introduces biobased polyurethanes and the nanomaterials that can be used as nanocomposites followed by the resulting polyurethane nanocomposites. The second part then explores important applications in paints and surface coatings, shape memory, self-healing, self-cleaning, biomaterials, and packaging materials. Written by a leading expert on polyurethane nanocomposites, the book is a great introduction to this smart material and its applications.

Inorganic Two-dimensional Nanomaterials: Fundamental Understanding, Characterizations, and Energy Applications

Editor: Changzheng Wu
ISBN: 978-1-78262-465-3, 2017

Synopsis

Inorganic 2D nanomaterials, or inorganic graphene analogs, are gaining great attention due to their unique properties and potential energy applications. They contain ultrathin nanosheet morphology with one-dimensional confinement, but unlike pure carbon graphene, inorganic two-dimensional nanomaterials have a more abundant elemental composition and can form different crystallographic structures. These properties contribute to their unique chemical reaction activity, tunable physical properties, and facilitate applications in the field of energy conversion and storage. **Inorganic Two-dimensional Nanomaterials** details the development of the nanostructures from computational simulation and theoretical understanding of their synthesis and characterization. Individual chapters then cover different applications of the materials as electrocatalysts, flexible supercapacitors, flexible lithium-ion batteries, and thermoelectrical devices. The book provides a comprehensive overview of the field for researchers working in the areas of materials chemistry, physics, energy, and catalysis.

Ionic Liquid Devices

Editor: Ali Eftekhari
ISBN: 978-1-78801-181-5, 2017

Synopsis

Ionic liquids are attractive because they offer versatility in the design of organic salts. As ion-rich media, ionic liquids can control the properties of the system by tuning the size, charge, and shape of the composing ions. Whilst the focus has mainly been on the potential applications of ionic liquids as solvents, they also provide innovative opportunities for designing new systems and devices. Limitations from the high viscosity and expensive purification of the ionic liquids are also not a barrier for applications as devices. Written by leading authors, **Ionic Liquid Devices** introduces the innovative applications of ionic liquids. Whilst the first chapters focus on their characterization, which can be difficult in some instances, the rest of the book demonstrates how ionic liquids can play substantial roles in quite different systems from sensors and actuators to biomedical applications. The book provides a comprehensive resource aimed at researchers and students in materials science, polymer science, chemistry, and physics interested in the materials and inspire the discovery of new applications of ionic liquids in smart devices.

Polymerized Ionic Liquids

Editor: Ali Eftekhari
ISBN: 978-1-78262-960-3, 2017

Synopsis

The applications of ionic liquids can be enormously expanded by arranging the organic ions in the form of a polymer architecture. Polymerized ionic liquids (PILs), also known as poly(ionic liquid)s or polymeric ionic liquids, provide almost all features of ionic polymers plus a rare versatility in design. The mechanical properties of the solid or solid-like polymers can also be controlled by external stimuli, the basis for designing smart materials. Known for over four decades, PILs is a member of the ionic polymers family. Although the previous forms of ionic polymers have a partial ionicity, PILs are entirely composed of ions. Therefore, they offer better flexibility for designing a responsive architecture as smart materials. Despite the terminology, PILs can be synthesized from solid organic ionic salts since the monomer liquidity is not a requirement for the polymerization process. Ionicity can also be induced to a neutral polymer by post-polymerization treatments. This is indeed an emerging field whose capabilities have been somehow overshadowed by the popularity of ionic liquids. However, recent reports in the literature have shown impressive potentials for the future. Written by leading authors, the present book provides a comprehensive overview of this exciting area, discussing various aspects of PILs and their applications as smart materials. Owing to the novelty of this area of research, the book will appeal to a broad readership including students and researchers from materials science, polymer science, chemistry, and physics.

Nanogels for Biomedical Applications

Editors: Arti Vashist, Ajeet K. Kaushik, Sharif Ahmad and Madhavan Nair
ISBN: 978-1-78262-862-0, 2017

Synopsis

Nanogel-based systems have gained tremendous attention due to their diverse range of applications in tissue engineering, regenerative medicine, biosensors, orthopedics, wound healing, and drug delivery. **Nanogels for Biomedical Applications** provides a comprehensive overview of nanogels and their use in nanomedicine. The book starts with the synthesis, methods, and characterization techniques for nanogel-based smart materials followed by individual chapters demonstrating the different uses of the materials. Applications covered include anticancer therapy, tuberculosis diagnosis and treatment, tissue engineering, gene delivery, and targeted drug delivery. The book will appeal to biologists, chemists, and nanotechnologists interested in translation research for personalized nanomedicine for health care.

Reactive Inkjet Printing: A Chemical Synthesis Tool

Editors: Patrick J Smith and Aoife Morrin
ISBN: 978-1-78262-767-8, 2017

Synopsis

Reactive inkjet printing uses an inkjet printer to dispense one or more reactants onto a substrate to generate a physical or chemical reaction to form a product in situ. Thus, unlike traditional inkjet printing, the printed film chemistry differs from that of the initial ink droplets. The appeal of reactive inkjet printing as a chemical synthesis tool is linked to its ability to produce droplets whose size is both controllable and predictable, which means that the individual droplets can be thought of as building blocks where droplets can be added to the substrate in a high precision format to give good control and predictability over the chemical reaction. The book starts by introducing the concept of using reactive inkjet printing as a building block for making materials. Aspects such as the behavior of printed droplets on the substrate and their mixing are discussed in the first chapters. The following chapters then discuss different applications of the technique in areas including additive manufacturing and silk production, production of materials used in solar cells, printed electronics, dentistry, and tissue engineering. Edited by two leading experts, **Reactive Inkjet Printing: A Chemical Synthesis Tool** provides a comprehensive overview of this technique and its use in fabricating functional materials for health and energy applications. The book will appeal to advanced level students in materials science.

Electrochromic Smart Materials: Fabrication and Applications

Editors: Jian Wei Xu, Ming Hui Chua and Kwok Wei Shah
ISBN: 978-1-78801-143-3, 2019

Synopsis

Interest in and attention on electrochromic technology has been growing since the 1970s, with the advent of numerous electrochromic devices in commercial and industrial settings. Many laboratory-based color-changing electrochromic device prototypes have surfaced following research breakthroughs in recent years, and the consumer market has been expanding continuously. Electrochromic devices have a wide range of applications, such as displays, self-dimming mirrors for automobiles, electrochromic e-skins, textiles, and smart windows for energy-efficient buildings. **Electrochromic Smart Materials** covers major topics related to the phenomenon of electrochromism, including fundamental principles, different classes, and subclasses of electrochromic materials, and device processing and manufacturing. It also highlights a broad range of existing and potential applications of electrochromic devices, with an analysis of the current market needs and future trends. Providing a comprehensive overview of the field, this book will serve as introductory reading to those new to this area, as well as a resource providing detailed, in-depth knowledge and insights to the seasoned audience. Featuring contributions from researchers across the globe, it will be of interest to postgraduate students and researchers in both academia and industry interested in smart design, materials science and engineering.

Layered Materials for Energy Storage and Conversion

Editors: Dongsheng Geng, Yuan Cheng and Gang Zhang
ISBN: 978-1-78801-426-7, 2019

Synopsis

The considerable interest in graphene and 2D materials is sparking intense research on layered materials due to their unexpected physical, electronic, chemical, and optical properties. This book will provide a comprehensive overview of the recent and state-of-the-art research progress on layered materials for energy storage and other applications. With a brief introduction to layered materials, the chapters of this book gather various fascinating topics such as electrocatalysis for fuel cells, lithium-ion batteries, sodium-ion batteries, photovoltaic devices, thermoelectric devices, supercapacitors, and water splitting. Unique aspects of layered materials in these fields, including novel synthesis and functionalization methods, particular physicochemical properties, and consequently enhanced performance are addressed. Challenges and perspectives for layered materials in these fields will also be presented. With contributions from key researchers, **Layered Materials for Energy Storage and Conversion** will be of interest to students, researchers, and engineers worldwide who want a basic overview of the latest progress and future directions.

Smart Membranes

Editor: Liang-Yin Chu
ISBN: 978-1-78801-243-0, 2019

Synopsis

Smart membranes that respond to environmental stimuli are gaining attention because of their potential use in a variety of applications, from drug delivery to water treatment. Their surface characteristics and/or permeation properties, including pressure-driven hydraulic permeability and concentration-driven diffusional permeability, can be adjusted in response to small chemical and/or physical stimuli in the environment. This book will cover topics such as novel design and fabrication strategies, approaches for controlling structure and performance, as well as cutting-edge applications of smart membranes. It will deliver new insights and fundamentals for both professionals and newcomers in related fields. Edited by an internationally renowned expert and with contributions from key researchers, **Smart Membranes** provides a comprehensive overview of the topic. It will appeal to students and researchers across materials science, chemistry, chemical engineering, pharmaceutical science, and biomedical science.

Cucurbituril-based Functional Materials

Editor: Dönüs Tuncel
ISBN: 978-1-78801-488-5, 2019

Synopsis

Smart materials constructed through supramolecular assemblies have been receiving considerable attention because of their potential applications, which include self-healing materials, energy storage, photonic devices, sensors and theranostics. Host–guest chemistry of various macrocyclic receptors with organic guests provides a unique way to control tailor-made nanoarchitectures for the formation of pre-designed functional materials. **Cucurbituril-based Functional Materials** provides an overview of this fascinating macrocycle, cucurbituril (CB) homologues and derivatives-based supramolecular nanostructured materials. Chapters cover the synthesis, properties and application of CB-based smart materials and nanostructures. With contributions from key researchers, this book will be of interest to students and researchers working in materials science, as well as those working on cucurbituril-based materials in organic and physical chemistry.

Subject Index